AF684485

PORSCHE SPORTS RACERS 1952-1968 Ultimate Portfolio

Compiled by R M Clarke

ISBN 185520 7303

BROOKLANDS BOOKS LTD.
P.O. BOX 146, COBHAM,
SURREY, KT11 1LG. UK
sales@brooklands-books.com

A-POSPUP

www.brooklands-books.com

Printed in China

Brooklands Books

MOTORING

ROAD TEST SERIES

Abarth Gold Portfolio 1950-1971
Alfa Romeo Giulietta Gold Portfolio 1954-1965
Alfa Romeo Giulia Coupés 1963-1976
Alfa Romeo Spider 1966-1990
Alfa Romeo Alfasud 1972-1984
Alfa Romeo Alfetta Gold Portfolio 1972-1987
Alfa Romeo Spider & GTV Perf. Port. 1995-2005
Alvis Gold Portfolio 1919-1967
AMC Rambler Limited Edition Extra 1956-1969
AMX & Javelin Gold Portfolio 1968-1974
Armstrong Siddeley Gold Portfolio 1945-1960
Aston Martin Gold Portfolio 1921-1947
Aston Martin Ultimate Portfolio 1948-1968
Aston Martin Ultimate Portfolio 1968-1980
Aston Martin Ultimate Portfolio 1981-1993
Aston Martin Ultimate Portfolio 1994-2006
Audi Quattro Gold Portfolio 1980-1991
Audi Quattro Takes On The Competition
Audi TT Performance Portfolio 1998-2006
Austin-Healey 100 & 100/6 Gold Port. 1952-1959
Austin-Healey 3000 Ultimate Portfolio 1959-1967
Austin-Healey Sprite Gold Portfolio 1958-1971
Bentley & Rolls-Royce Portfolio 1990-2002
Berkeley Sportscars Limited Edition
BMW 6 & 8 Cyl. Cars Limited Edition 1935-1960
BMW 700 Limited Edition 1959-1965
BMW 1600 Collection No. 1 1966-1981
BMW 2002 Ultimate Portfolio
BMW 6 Cylinder Coupés & Saloons Gold P. 1969-1976
BMW 316, 318, 320 (4 cyl.) Gold Port. 1975-1990
BMW 320, 323, 325 (6 cyl.) Gold Port. 1977-1990
BMW 3 Series Gold Portfolio 1991-1997
BMW 5 Series Gold Portfolio 1988-1995
BMW 6 Series Ultimate Portfolio 1976-1989
BMW 7 Series Performance Portfolio 1977-1986
BMW 7 Series Performance Portfolio 1986-1993
BMW 8 Series Performance Portfolio
BMW Alpina Performance Portfolio 1967-1987
BMW Alpina Performance Portfolio 1988-1998
BMW Z3, M Coupe & M Roadster Gold Port. 1996-02
Borgward Isabella Limited Edition
Bristol Cars Portfolio
Buick Performance Portfolio 1947-1962
Buick Muscle Portfolio 1963-1973
Buick Riviera Performance Portfolio 1963-1978
Cadillac Performance Portfolio 1948-1958
Cadillac Performance Portfolio 1959-1966
Cadillac Eldorado Performance Portfolio 1967-1978
Cadillac Allante Limited Edition Extra
Chevrolet 1955-1957
Impala & SS Muscle Portfolio 1958-1972
Corvair Performance Portfolio 1959-1969
El Camino & SS Muscle Portfolio 1959-1987
Chevy II & Nova SS Gold Portfolio 1962-1974
Chevelle & SS Gold Portfolio 1964-1972
Camaro Muscle Portfolio 1967-1973
Camaro Performance Portfolio 1993-2000
Chevrolet Corvette Gold Portfolio 1953-1962
Chevrolet Corvette Sting Ray Gold Port. 1963-1967
Chevrolet Corvette Gold Portfolio 1968-1977
High Performance Corvettes 1983-1989
Chrysler Imperial Gold Portfolio 1951-1975
Valiant 1960-1962
PT Cruiser Performance Portfolio
Citroen 2CV Ultimate Portfolio 1948-1990
Citroen DS & ID 1955-1975
Citroen DS & ID Gold Portfolio 1955-1975
Citroen SM Limited Edition Extra 1970-1975
Shelby Cobra Gold Portfolio 1962-1969
Crosley & Crosley Specials Limited Edition
Cunningham Automobiles 1951-1955
Datsun Roadsters Performance Portfolio 1960-71
Datsun 240Z & 260Z Gold Portfolio 1970-1978
DeLorean Gold Portfolio 1977-1995
De Soto Limited Edition 1952-1960
Dodge Limited Edition 1949-1959
Dodge Dart Limited Edition Extra 1960-1976
Dodge Muscle Portfolio 1964-1971
Charger Muscle Portfolio 1966-1974
ERA Gold Portfolio 1934-1994
Facel Vega Limited Edition Extra 1954-1964
Ferrari Limited Edition 1947-1957
Ferrari Limited Edition 1958-1963
Ferrari Dino Limited Edition Extra 1965-1974
Ferrari 328 348 Mondial Ultimate Portfolio 1986-94
Ferrari F355 & 360 Gold Portfolio 1995-2004
Fiat 600 & 850 Gold Portfolio 1955-1972
Fiat Dino Limited Edition
Fiat X1/9 Gold Portfolio 1973-1989
Ford Consul, Zephyr, Zodiac Mk. I & II 1950-1962
Ford Zephyr, Zodiac, Executive Mk. III & IV 1962-1971
High Performance Capris Gold Portfolio 1969-1987
Capri Muscle Portfolio 1974-1987
High Performance Fiestas 1979-1991
Ford Escort RS & Mexico Limited Edition 1970-1979
High Performance Escorts Mk. II 1975-1980
High Performance Escorts 1980-1985
High Performance Escorts 1985-1990
Ford Thunderbird Performance Portfolio 1955-1957
Ford Thunderbird Performance Portfolio 1958-1963
Ford Thunderbird Performance Portfolio 1964-1976
Ford Fairlane Performance Portfolio 1955-1970
Ford Ranchero Muscle Portfolio 1957-1979
Edsel Limited Edition 1957-1960
Ford Galaxie & LTD Gold Portfolio 1960-1976
Falcon Performance Portfolio 1960-1970
Ford Torino Performance Portfolio 1968-1974
Ford Bronco 4x4 Performance Portfolio 1966-1977
Ford Bronco 1978-1988
Shelby Mustang Ultimate Portfolio 1965-1970
Mustang Muscle Portfolio 1967-1973
High Performance Mustang IIs 1974-1978
Mustang 5.0L Muscle Portfolio 1982-1993
Mustang 5.0L Takes On The Competition
Goggomobil Limited Edition
Holden 1948-1962
Honda S500 • S600 • S800 Limited Edition 1962-1970
Honda CRX 1983-1987
Hudson Performance Portfolio 1946-1957
International Scout Gold Portfolio 1961-1980
Isetta Gold Portfolio 1953-1964
Jaguar and SS Gold Portfolio 1931-1951
Jaguar C-Type & D-Type Gold Portfolio 1951-1960
Jaguar XK120, 140, 150 Gold Portfolio 1948-1960
Jaguar Mk. VII, VIII, IX, X, 420 Gold Port. 1950-1970
Jaguar E-Type Gold Portfolio 1961-1971
Jaguar XJ6 Series I & II Gold Portfolio 1968-1979
Jaguar XJ6 Series III Perf. Portfolio 1979-1986
Jaguar XJS Gold Portfolio 1975-1988
Jaguar XJ-S V12 Ultimate Portfolio 1988-1996
Jaguar XK8 Limited Edition
Jeep CJ-5 Limited Edition 1960-1975
Jeep CJ-5 & CJ-7 4x4 Perf. Portfolio 1976-1986
Jeep Wagoneer Performance Portfolio 1963-1991
Jeep J-Series Pickups 1970-1982
Jeepster & Commando Limited Edition 1967-1973
Jeep Cherokee & Comanche Pickups P. P. 1984-91
Jeep Wrangler 4x4 Performance Portfolio 1987-99
Jeep Cherokee & Grand Cherokee 4x4 P. P. 1992-98
Jensen Interceptor Ultimate Portfolio 1966-1992
Jensen - Healey Limited Edition 1972-1976
Kaiser - Frazer Limited Edition 1946-1955
Lagonda Gold Portfolio 1919-1964
Lancia Aurelia & Flaminia Gold Portfolio 1950-1970
Lancia Fulvia Gold Portfolio 1963-1976
Lancia Beta Gold Portfolio 1972-1984
Lancia Stratos Limited Edition Extra
Lancia Delta & integrale Ultimate Portfolio
Land Rover Series I, II & IIA Gold Portfolio 1948-71
Land Rover Series III 4x4 Perf. Portfolio 1971-1985
Land Rover 90 110 Defender Gold Portfolio 1983-94
Land Rover Discovery Perf. Port. 1989-2000
Fifty Years of Selling Land Rover
Lamboghini Performance Portfolio 1964-1976
Lamboghini Performance Portfolio 1977-1989
Lamboghini Gold Portfolio 1990-2004
Lincoln Gold Portfolio 1949-1960
Lincoln Continental Performance Portfolio 1961-1969
Lincoln Continental 1969-1976
Lotus Sports Racers Portfolio - covering 1951-1965
Lotus Seven Gold Portfolio 1957-1973
Lotus Elite Limited Edition 1957-1964
Lotus Elan Ultimate Portfolio 1962-1974
Lotus Elan & SE 1989-1992
Lotus Europa Gold Portfolio 1966-1975
Lotus Elite & Eclat 1974-1982
Lotus Elise & Exige Gold Portfolio 1995-2005
Marcos Coupés & Spyders Gold Portfolio 1960-1997
Maserati Cars Performance Portfolio 1957-1970
Maserati Cars Performance Portfolio 1971-1982
Maserati Cars Performance Portfolio 1982-1998
Matra Limited Edition 1965-1983
Mazda Miata MX-5 Performance Portfolio 1989-1997
Mazda Miata MX-5 Performance Portfolio 1998-2005
Mazda Miata MX-5 Takes On The Competition
Mazda RX-7 Gold Portfolio 1978-1991
McLaren F1 • GTR • LM Sportscar Perf. Portfolio
Mercedes 190 & 300 SL 1954-1963
Mercedes S & 600 Limited Edition Extra 1965-1972
Mercedes S Class 1972-1979
Mercedes S Class Limited Edition Extra 1980-1991
Mercedes 230 • 250 • 280SL Gold Portfolio 1963-1971
Mercedes-Benz SLs & SLCs Ultimate Port. 1971-89
Mercedes SLs Performance Portfolio 1989-1994
Mercedes G-Wagen Gold Portfolio 1981-2005
Mercedes 190 Limited Edition Extra 1983-1993
Mercedes CLK & SLK Limited Edition
Mercury Gold Portfolio 1947-1966
Mercury Comet & Cyclone Lim. Edit. Extra 1960-75
Cougar Muscle Portfolio 1967-1973
Messerschmitt Gold Portfolio 1954-1964
MG Gold Portfolio 1929-1939
MG TA & TC Gold Portfolio 1936-1949
MG TD & TF Gold Portfolio 1949-1955
MGA & Twin Cam Gold Portfolio 1955-1962
MG Midget Gold Portfolio 1961-1979
MGB Roadsters 1962-1980
MGB MGC & V8 Gold Portfolio 1962-1980
MGC & MGB GT V8 Limited Edition
MGF & TF Performance Portfolio 1995-2005
Mini Gold Portfolio 1959-1969
Mini Gold Portfolio 1969-1980
Mini Gold Portfolio 1981-1997
High Performance Minis Gold Portfolio 1960-1973
Mini Cooper Gold Portfolio 1961-1971
Mini Moke Ultimate Portfolio 1964-1994
Mini Performance Portfolio 2001-2006
Starion & Conquest Performance Portfolio 1982-90
Mitsubishi 3000GT & Dodge Stealth P.P. 1990-99
Morgan Three-Wheeler Gold Portfolio 1910-1952
Morgan Plus 4 & Four 4 Gold Portfolio 1936-1967
Morgan Cars Portfolio 1968-2001
Morris Minor Collection No. 1 1948-1980
Nash Limited Edition Extra 1949-1957
Nash-Austin Metropolitan Gold Portfolio 1954-1962
Nissan Skyliner GT-R Limited Edition Extra 1989-02
NSU Ro80 Limited Edition
NSX Performance Portfolio 1989-1999
Oldsmobile Automobiles 1955-1963
Oldsmobile Muscle Portfolio 1964-1971
Cutlass & 4-4-2 Muscle Portfolio 1964-1974
Opel GT Ultimate Portfolio 1968-1973
Opel Manta Limited Edition 1970-1975
Pantera Ultimate Portfolio 1970-1995
Panther Gold Portfolio 1972-1990
Plymouth Limited Edition 1950-1960
Plymouth Fury Limited Edition Extra 1956-1976
Barracuda Muscle Portfolio 1964-1974
Plymouth Muscle Portfolio 1964-1971
High Performance Firebirds 1982-1988
Firebird & Trans Am Performance Portfolio 1993-00
Pontiac Fiero Performance Portfolio 1984-1988
Porsche Sports Racers Ultimate Portfolio 1952-1968
Porsche 917 • 935 • 956 • 962 Sports Racers Gold Port.
Porsche 912 Limited Edition Extra
Porsche 365 Ultimate Portfolio 1952-1965
Porsche 911 1965-1969
Porsche 911 1973-1977
Porsche 911 SC & Turbo Gold Portfolio 1978-1983
Porsche 911 Carrera & Turbo Gold Port. 1984-1989
Porsche 911 Ultimate Portfolio 1990-1997
Porsche 911 Takes On The Competition 1990-1997
Porsche 911 Ultimate Portfolio 1998-2004
Porsche 914 Ultimate Portfolio
Porsche 924 Gold Portfolio 1975-1988
Porsche 928 Gold Portfolio 1977-1995
Porsche 928 Takes On The Competition
Porsche 944 Ultimate Portfolio
Porsche 968 Limited Edition Extra
Porsche Boxster Ultimate Portfolio 1996-2004
Railton & Brough Superior Gold Portfolio 1933-1950
Range Rover Gold Portfolio 1970-1985
Range Rover Gold Portfolio 1985-1995
Range Rover Performance Portfolio 1995-2001
Range Rover Takes on the Competition
Riley Gold Portfolio 1924-1939
Rolls-Royce Silver Clould & Bentley S Ultimate Port.
Rolls-Royce Silver Shadow Ultimate Port. 1965-80
Rolls-Royce & Bentley Gold Portfolio 1980-1989
Rover P4 1949-1959
Rover 2000 & 2200 1963-1977
Subaru Impreza Turbo Limited Edition Extra 94-00
Subaru Impreza WRX Performance Port. 2001-05
Studebaker Hawks & Larks 1956-1963
Avanti Limited Edition Extra 1962-1991
Sunbeam Alpine Limited Edition Extra 1959-1968
Sunbeam Tiger Limited Edition Extra 1964-1967
Suzuki SJ Gold Portfolio 1971-1997
Vitara, Sidekick & Geo Tracker Perf. Port. 1988-1997
Toyota Land Cruiser Gold Portfolio 1956-1987
Toyota Land Cruiser 1988-1997
Toyota Supra Performance Portfolio 1982-1998
Toyota MR2 Gold Portfolio 1984-1997
Toyota MR2 Takes On The Competition
Triumph TR2 & TR3 Gold Portfolio 1952-1961
Triumph TR4, TR5, TR250 1961-1968
Triumph TR6 Gold Portfolio 1969-1976
Triumph Herald 1959-1971
Triumph Vitesse 1962-1971
Triumph Spitfire Gold Portfolio 1962-1980
Triumph 2000, 2.5, 2500 1963-1977
Triumph GT6 Gold Portfolio 1966-1974
Triumph Stag Gold Portfolio 1970-1977
Triumph Dolomite Sprint Limited Edition
TVR Gold Portfolio 1959-1986
TVR Performance Portfolio 1986-1994
TVR Performance Portfolio 1995-2000
TVR Performance Portfolio 2000-2005
VW Beetle Gold Portfolio 1935-1967
VW Beetle Gold Portfolio 1968-1991
VW Bus, Camper, Van Perf. Portfolio 1954-1967
VW Bus, Camper, Van Perf. Portfolio 1968-1979
VW Bus, Camper, Van Perf. Portfolio 1979-1991
VW Karmann Ghia Gold Portfolio 1955-1974
VW Scirocco 1974-1981
VW Golf GTI Limited Edition Extra 1976-1991
VW Corrado Limited Edition 1989-1995
Volvo PV444 & PV544 Perf. Portfolio 1945-1965
Volvo 120 Amazon Ultimate Portfolio
Volvo 1800 Ultimate Portfolio 1960-1973
Volvo 140 & 160 Series Gold Portfolio 1966-1975
Forty Years of Selling Volvo
Westfield Performance Portfolio 1982-2004

CAR AND DRIVER SERIES

Car and Driver on BMW 1957-1977
Car and Driver on Corvette 1978-1982
Car and Driver on Corvette 1983-1988
Car and Driver on Ferrari 1955-1962
Car and Driver on Ferrari 1963-1975
Car and Driver on Mopar 1956-1967
Car and Driver on Mustang 1964-1973
Car and Driver on Porsche 1955-1962
Car and Driver on Porsche 1963-1970
Car and Driver on Porsche 1970-1976
Car and Driver on Porsche 1977-1981
Car and Driver on Porsche 1982-1986

RACING & THE LAND SPEED RECORD

The Land Speed Record 1898-1919
The Land Speed Record 1920-1929
The Land Speed Record 1930-1939
The Land Speed Record 1940-1962
The Land Speed Record 1963-1999
Can-Am Racing 1966-1969
Can-Am Racing 1970-1974
Can-Am Racing Cars 1966-1974
The Carrera Panamericana Mexico - 1950-1954
Le Mans - The Bentley & Alfa Years - 1923-1939
Le Mans - The Jaguar Years - 1949-1957
Le Mans - The Ferrari Years - 1958-1965
Le Mans - The Ford & Matra Years - 1966-1974
Le Mans - The Porsche Years - 1975-1982
Le Mans - The Porsche & Jaguar Years - 1983-91
Le Mans - The Porsche & Peugeot Years - 1992-99
Mille Miglia - The Alfa & Ferrari Years - 1927-1951
Mille Miglia - The Ferrari & Mercedes Years - 1952-57
Targa Florio - The Porsche & Ferrari Years - 1955-1964
Targa Florio - The Porsche Years - 1965-1973

ROAD & TRACK 'PORTFOLIO' SERIES

Road & Track BMW M Series Portfolio 1979-2002
R & T BMW Z3, M Coupe & M Roadster Port. 96-02
R & T Camaro & Firebird Portfolio 1993-2002
Road & Track Corvette Portfolio 1997-2002
Road & Track Dodge Viper Portfolio 1992-2002
Road & Track Ferrari V-12 Portfolio 1992-2002
Road & Track Ferrari F355 360 F430 Portfolio 95-06
R & T Jaguar XJ-S - XK8 - XKR Portfolio 1975-2003
R & T Mercedes SL - SLK - CLK Portfolio 1990-03
Road & Track MX-5 Miata Portfolio 1989-2002
Road & Track Mustang Portfolio 1994-2002
Road & Track Nissan 300ZX & 350Z Portfolio 1984-03
Road & Track Porsche 928 Portfolio 1977-1994
Road & Track Porsche 911 Portfolio 1990-1997

ROAD & TRACK 'ON' SERIES

Road & Track on Aston Martin 1962-1990
Road & Track on Audi & Auto Union 1952-1980
Road & Track on Audi & Auto Union 1980-1986
Road & Track on Austin Healey 1953-1970
Road & Track on BMW Cars 1966-1974
Road & Track on BMW Cars 1975-1978
Road & Track on BMW Cars 1979-1983
R & T on Cobra, Shelby & Ford GT40 1962-1992
Road & Track on Corvette 1968-1982
Road & Track on Corvette 1982-1986
Road & Track on Corvette 1986-1990
Road & Track on Ferrari 1975-1981
Road & Track on Ferrari 1981-1984
Road & Track on Ferrari 1984-1988
Road & Track on Fiat Sports Cars 1968-1987
Road & Track on Jaguar 1950-1960
Road & Track on Jaguar 1961-1968
Road & Track on Jaguar 1968-1974
Road & Track on Jaguar 1974-1982
Road & Track on Jaguar 1983-1989
Road & Track on Lamborghini 1964-1985
Road & Track on Mercedes 1952-1962
Road & Track on Mercedes 1963-1970
Road & Track on Mercedes 1971-1979
Road & Track on MG Sports Cars 1949-1961
Road & Track on MG Sports Cars 1962-1980
Road & Track on Pontiac 1960-1983
Road & Track on Porsche 1951-1967
Road & Track on Porsche 1968-1971
Road & Track on Porsche 1972-1975
Road & Track on Porsche 1975-1978
Road & Track on Porsche 1979-1982
Road & Track on Porsche 1985-1988
R & T on Rolls Royce & Bentley 1950-1965
R & T on Rolls Royce & Bentley 1966-1984
Road & Track on Saab 1972-1992
R & T on Toyota Sports & GT Cars 1966-1984
R & T on Triumph Sports Cars 1967-1974
R & T on Triumph Sports Cars 1974-1982
Road & Track on Volkswagen 1951-1968
Road & Track on Volkswagen 1968-1978
Road & Track on Volkswagen 1978-1985
Road & Track on Volvo 1957-1974
Road & Track on Volvo 1977-1994

ROAD & TRACK BY AUTHOR

Road & Track - Henry Manney at Large & Abroad
Road & Track - Best of PS
Road & Track - Peter Egan Side Glances 1983-92
Road & Track - Peter Egan Side Glances 1992-97
Road & Track - Peter Egan Side Glances 1998-02
Road & Track - Peter Egan Side Glances 2002-06

PRACTICAL CLASSICS SERIES

PC on Midget/Sprite Restoration
PC on MGB Restoration
PC on Sunbeam Rapier Restoration

HOT ROD 'ENGINE' SERIES

Chevy 265 & 283
Chevy 302 & 327
Chevy 348 & 409
Chevy 396 & 427
Chevy 454 thru 512
Chrysler Hemi
Chrysler 273, 318, 340 & 360
Chrysler 361, 383, 400, 413, 426 & 440
Ford 289, 302, Boss 302 & 351W
Ford 351C & Boss 351
Ford Big Block

RESTORATION & GUIDE SERIES

BMW 2002 - A Comprehensive Guide
BMW '02 Restoration Guide
BMW E30 - 3 Series Restoration Bible
Classic Camaro Restoration
Chevrolet High Performance Tips & Techniques
Engine Swapping Tips & Techniques
Land Rover Restoration Portfolio
Lotus Elan Restoration Guide
MG 'T' Series Restoration Guide
MGA Restoration Guide
Mustang Restoration Tips & Techniques
Practical Gas Flow
Restoring Sprites & Midgets an Enthusiast's Guide
SU Carburetters Tuning Tips & Techniques
The Great Classic Muscle Cars Compared

MILITARY VEHICLES

Complete WW2 Military Jeep Manual
Dodge WW2 Military Portfolio 1940-1945
German Military Equipment WW2
Hail To The Jeep
Combat Land Rover Portfolio No. 1
Land Rover Military Portfolio
Military & Civilian Amphibians 1940-1990
Off Road Jeeps Civilian & Military 1944-1971
US Military Vehicles 1941-1945
Standard Military Motor Vehicles-TM9-2800 (WW2)
VW Kubelwagen Military Portfolio 1940-1990
WW2 Allied Vehicles Military Portfolio 1939-1945
WW2 Jeep Military Portfolio 1941-1945

MOTORCYCLES

To see our range of over 60 titles visit
www.brooklands-books.com

20/07Z6

Brooklands Books

CONTENTS

CONTENTS - *continued*

ACKNOWLEDGEMENTS

Brooklands Books has been publishing compilation racing and road test books on a wide range of makes and models for over 50 years and today is the world's largest publisher of such material. The company is grateful to the enthusiasts who support us and to the publishers of the original material who graciously allow us to include their copyright material. Our thanks in this instance go to the managements of *911 & Porsche World, Australian Motor Sports, Auto, Auto Topics, Autocar, Autocourse, Autosport, Car, Car and Driver, Car South Africa, Classic & Sportscar, Classic Cars, Foreign Car Guide, Motor, Motor Manual, Motor Sport, Motor Trend, Road & Track, Speed Age, Sporting Motorist, Sports Car Graphic, Sports Cars Illustrated, Wheels, World's Fastest Sports Cars* and the *World Car Catalogue.*

R.M. Clarke

Porsche has always marched to the beat of its own drum and rarely paid attention to the activities of others. This was never more evident than in their range of sports racing cars starting with the Type 550 in 1953. Designed for competition purposes, it featured a complex flat four (a so-called "boxer") that produced 110bhp from just 1.5 litres. Imagine that in 1953—more than half a century ago! Not only was it light, it was very quick and soon became a class winner in major long distance races. It also became quite a collector's piece.

For much of the subsequent sixty years Porsche's sports and racing cars have been powered by boxer engines of varying cylinder count and complexity, from flat fours all the way up to a monster flat-16. The most charismatic of the company's sports racing cars were, in the opinion of many enthusiasts, those from the period that embraced the Type 550 and its subsequent developments through to the RS61.

A new era was ushered in with the release of the stunning Type 904 that was shown to the media towards the end of 1963 and entered competition the following year. The 904 was powered by a 2-litre development of the quad cam boxer engine that was first seen in the 550. Engine apart, the 904 was a new ball game: fibreglass body of exquisite beauty, and a totally new chassis with coil spring suspension. It was to have a long and successful racing career and spawned more great racing Porsches over the next decade beginning with the fabulous Carrera 6, the first Porsche to be powered by a flat six engine.

The Porsche sports racing cars from this era were that rare combination of athleticism and beauty. They were classically elegant in their proportions, were technically interesting, very powerful for their engine size, very quick and extremely reliable. They were also versatile, particularly the 904 that won countless races on the track as well as many rallies.

Thankfully many of the small numbers of these cars that were built have survived and are often seen in classic events around the world. And they still look and sound great.

Gavin Farmer

montlhéry enduro

1100 cc Porsche Coupe Records

	OLD RECORD	NEW RECORD
500 Miles	98.25 mph	100.15 mph
6 Hours	98.90 "	100.85 "
1,000 Kilometers	99.15 "	101.10 "

1500 cc Porsche Coupe Records

	OLD RECORD	NEW RECORD
3,000 Kilometers	95.25 mph	98.50 mph
2,000 Miles	95.40 "	98.65 "
24 Hours	94.40 "	98.60 "
4,000 Kilometers	94.40 "	98.65 "
3,000 Miles	91.52 "	98.75 "
5,000 Kilometers	91.48 "	98.80 "
4,000 Miles	91.75 "	98.00 "
48 Hours	91.90 "	97.10 "
5,000 Miles	89.60 "	97.00 "
10,000 Kilometers	79.25 "	95.75 "
72 Hours	90.48 "	94.60 "

1500 cc Porsche-Glockler Records

	OLD RECORD	NEW RECORD
500 Kilometers	114.90 mph	116.60 mph
1,000 Kilometers	113.20 "	115.30 "
6 Hours	112.25 "	114.35 "

17 new international records established by porsche

A section of the course as seen from the driver's seat. The strange angle is a result of the car's position high on a banked turn

Photos by Kurt Wörner

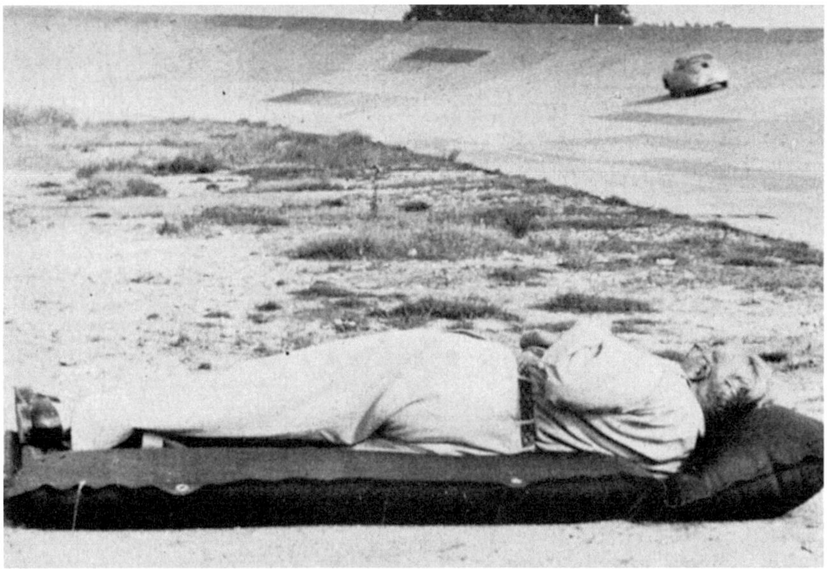

One of the drivers (left) takes a bit of well-earned "sack time" as the little coupe buzzes by behind him

The special-bodied Porsche-Glöckler (below left) had all four tires changed, refueled, replaced drivers, and was completely checked over in less than two minutes

A Porsche crewmember gives the 10,000 km. sign (above). Drivers Glöckler, Mueller, von Hanstein, Ramelow, and von Frankenberg made so many laps of the Montlhéry course that they were never quite sure what their total mileage was

The well-proven Volkswagen-based engine as fitted to the record-breaking cars. The large shroud houses the turbine cooling fan

Nurburgring . . . . Trenkel at speed in modified Porsche.

Nurburgring . . . . Glochler in his new Glochler-Porsche.

Porsche . . .

Racing

⬆ Dieburg, Germany . . . Six Porsche coupes line up. They dominate 1100 and 1500 cc racing in Germany.

⬇ Montlhery, France . . . one of three record-breaking Porsches which took 17 International records and one World's record. The 1100 cc coupe averaged 100.72 mph for 1000 kilometers (621.4 miles); the 1500 cc coupe set 11 records, among them 3000 miles at 98.95 mph and 72 hours at 94.66 mph; the Glochler-Porsche set three records including 500 kilometers (310.69 miles) at 116.88 mph and six hours at 114.74 mph.

Vero Beach, Florida . . . Max Hoffman at speed in his Glochler-Porsche. He placed first in the 1½ litre class and made second best time of the day.

Porsche trio . . . Lauprecht leads Count Einsiedel thru Dieburg esses.

Le Mans, France . . . competition coupe, which features an all-aluminum body and other modifications, won the 1100 cc class in 1951 and 1952.

Liege to Rome and Back

The winning Porsche in the Alps during the exhausting 3230 mile long distance run.

Porsche triumphed in one of Europe's most famous rallies, run over 3230 miles where drivers face the added handicap of everyday traffic.

Rugged mountain driving took the greatest toll of entrants. Only 24 cars were able to finish out of 106 starters.

Text and Photos

By GUNTHER F. MOLTER

The International Liége-Rome-Liége Rally is the famous long distance race on open roads in Europe. Since the roads are not blockaded, contestants face the added handicap of coping with everyday traffic.

The most recent rally was one of the most difficult. In less than 90 hours the drivers had to cover 3230 miles across Belgium, France, Italy and Germany. The fact that of 106 cars, which started from Liége, only 24 reached the finish line tells the story.

The route crossed the Alps and included 33 mountain passes, 16 at more than 6000 feet.

The winning car was a 1500 CC (91 cubic inch) Porsche 356 SL, lent to the drivers—two Germans, Helmut Polensky and Walter Schluter—by Miss Gilberte Thirion, a Belgian race driver and holder of the flying kilometer (0.6214 mile) record for a Porsche in Europe. In addition to the efforts of victors, the entire Porsche team completed the race. ☆ ☆

The aerodynamic body of the rear-engined two-seater Porsche offers little resistance to the airstream. The external petrol filler breaks the smooth contour. The underside of the body has been designed to ensure a smooth flow of air.

POTENT PORSCHES

Open Two-seater Type 550 : Modified 1.3-litre Model

THE new Porsche Type 550 two-seater sports model is now in production after being much in the news since its introduction at the 1953 Paris Show. Its outstanding performance in the Pan-American road race (in which it won the sports car class for cars up to 1,600 c.c. at a speed of 79.9 m.p.h.) did much to enhance its reputation in the prototype stage. This extremely fleet-looking sports car originally carried the standard Porsche engine which has inclined valves operated by push rods. Several examples, thus equipped, scored numerous successes in German events during the past year and this led to the development of the Type 550 power unit which has twin overhead camshafts.

The new engine has a capacity of 1,498 c.c. and develops 110 b.h.p. at 7,000 r.p.m. The exhaust valves are sodium cooled and each camshaft drives a separate distributor, both of which lie in a horizontal position on either side of the power unit. Lubrication is effected by a dry sump system. A dry single-plate clutch transmits power to the rear wheels through a four-speed gear box which has synchromesh on all gears. The air-cooled, rear engine retains the basic characteristics of the Volkswagen power unit from which it was originally developed.

Twin-choke Solex carburettors are fitted, one to each bank of cylinders, which are cast in light alloy. Each individual cylinder has two sparking plugs. The fuel tank and oil radiator are located in the nose of the car to counteract any tendency to tail heaviness; the capacity of the fuel tank is 15 gallons. One of the most interesting features of this new engine is the twin turbine which is fitted to supplement the cooling of the cylinder barrels; cool air is drawn through an intake at the back of the car (see photograph on left) and carried to each bank of cylinders

The body has been subjected to a great deal of research and it is entirely built up of alloy panels which have been welded and riveted to ensure a thoroughly rigid structure. The underside is so designed that an undisturbed flow of air is attained. Two separate seats are fitted and a fabric hood is supplied. The centre lock wheels are shod with 5.00-16in tyres. The manufacturers claim that the total weight of the car is slightly under 12cwt and the maximum speed is said to be in excess of 140 m.p.h.

In addition to the standard Porsche models the company last year introduced a 1.5-litre model which is known as the Super. For 1954 it is intended to increase this range and manufacture a 1.3-litre Super. Two protoypes of this proposed model were entered in the 1,000-kilometre race at Nurburg Ring last August and both cars passed the test admirably. The main difference in the modified engine of this type is that the crankshaft is specially built up to withstand sustained high speeds and the main and big-end bearings are supported by roller bearings. Production is expected to begin shortly

The new air-cooled flat four engine, based on that of the Volkswagen, develops 110 b.h.p. at 7,000 r.p.m., Interesting features include twin overhead camshafts, a twin-choke Solex carburettor for each bank of cylinders and two sparking plugs for each cylinder.

OVER TO YOU, Herr Bolster, says Herr Jeser, taking the passenger's seat in the Spyder 550 Porsche with which he finished second to Olivier's similar car in the Paris 24 Hours race.

JOHN BOLSTER TRIES

THE PORSCHE 550 SPYDER

—which finished 2nd in the Paris 24 Hours Race

WHEN the race for the Vingt-Quatre Heures de Paris was one hour old, I met my friend, Serge Pozzoli, and shook hands all continental. We watched the Gordinis and the Maseratis screaming round the lip of the Montlhéry banking and—"Formidable!" said Serge, and "Jolly good show," said *L'anglais Bolstaire*.

Whereupon, we both announced that a Porsche victory was certain, and so it proved. The little four-camshaft Spyders were first and second, their third entry having been involved in one of the many crashes. As the 2-litre class dwindled and died, the Porsches proved, once again, that they are 24-hour cars.

On the Tuesday after the race, I had arranged to do some testing at Montlhéry, and so my friend, "Jabby" Crombac, approached the winners and asked if I could also try one of their cars. They said at once that they would bring both machines along, but apologized because they could not be washed and polished, as the mechanics were naturally resting. Anyway, the two tiny, low "beetles" duly arrived, and I was allowed to examine them.

The general chassis details of the Spyder follow normal Porsche practice. There is independent suspension all round by torsion bars, the front wheels pivoting on trailing arms. At the rear there are swing axles, reinforced by trailing radius arms. The engine location differs entirely from that of the everyday cars, however.

In the case of the Porsches I have driven before, the power unit lives right out in the tail. The competition model has the engine forward of the swing axles, and only the gearbox is behind. This gives a more suitable weight distribution for phenomenally high speeds in a very light car.

The engine is still an air-cooled flat four, but it is specially built for racing. Four overhead camshafts operate the inclined valves, and there is an oil cooler which gets its air from a slot in the nose. This has a shutter controlled by the driver for temperature regulation. The whole unit is proportioned for continuous operation at high revolutions. In winning the 24-hour race at Montlhéry, both cars were run throughout up to 7,000 r.p.m., but 7,500 r.p.m. is the normal speed for shorter races. 110 b.h.p. is the maintained power output of this ultra-efficient $1\frac{1}{2}$-litre unit.

The well-known Porsche gearbox, with synchromesh on all four speeds, is employed. It is an all-indirect box, with closer ratios than standard. The gear lever is in the centre of the car, being operated by the right hand as the steering is on the left.

The arrival of the Porsches at the track was heralded by wind and rain, but they were soon out on the concrete and lapping steadily. After a few warming up laps at around 110 m.p.h. the throttles were opened, and a number of tours were put in at 122 m.p.h. The little machines seemed remarkably steady in spite of the extremely unfavourable weather conditions.

The cockpit cover of No. 51, Herr Jeser's car, was then removed, and I occupied the passenger seat. As there was no windscreen my side, I accepted an unmerciful buffeting, but it was quite an experience being whistled round the banking. I am not a good passenger, but it was obvious that the car was stable, and there was nothing alarming or spectacular about the ride.

FLAT - FOUR air-cooled 4 o.h.v. engine (above) of the Spyder is set ahead of the rear swing axles and aft of the driver.

★

FLAT OUT: (Right) John Bolster, with Jeser passengering, passes the Montlhéry stands at around 125 m.p.h. with the amazing $1\frac{1}{2}$-litre Porsche.

Subsequently, we changed places. Bottom gear is very high, but the smooth clutch looks after that, and one is soon away. The gearbox is a sheer delight, with a light, positive action and very close ratios. I was well past 100 m.p.h. when a flick of two fingers gave me top speed.

The engine had plenty of power in the middle ranges and was quite flexible. It was smooth and quiet, feeling entirely happy to maintain the 7,000 r.p.m. that it had already sustained for 24 hours. On the drying track, I at once allowed the speedometer needle to run past the 200 k.p.h. mark, which is around 125 m.p.h.

I can honestly say that there is no feeling of having a rear engine when one drives this car. In spite of its light weight, it is easier to drive than the normal production Porsche, which needs a little practice before one is fully at home with it. Naturally, the central mounting of the engine pushes the driver and passenger very far forward. The driving position is reminiscent of a "500", and the very short bonnet is forgotten after a lap or two. The brakes had not suffered in spite of a day and a night of road racing and, of course, they are as powerful as would be expected.

The Porsche 550 Spyder is a superbly constructed and extremely fast $1\frac{1}{2}$-litre competition car. I can think of no vehicle I would rather drive if I wanted to be sure of finishing in any long distance race.

Compact build allows room for 25-gal. gas tank, 3-gal. oil tank.

The "550" engine: 1) cooling blower 2) breather 3) l.h. distributor 4) flywheel 5) Hirth-type built-up crankshaft 6) subshaft drive wheel 7) crankcase 8) generator 9) cooling fan 10) tach drive 11) one-piece roller bearing big end 12) pair of pinions driving transverse shafts 13) oil pressure warning light switch 14) oil pump.

A VISIT TO PORSCHE:

THE SPYDER TYPE 550

by Hansjoerg Bendel

WITH the German Motor Show at Frankfurt coming up, it was fairly safe to assume that Porsche would be preparing something new, and so I arranged a visit to their works situated at Zuffenhausen near Stuttgart. I was received by Porsche-Racing Manager Huschke von Hanstein, winner of the shortened 1940 Mille Miglia, and Hans Klauser, who is responsible for the efficient technical synchronization of the various departments within the factory, and thereby is in close touch with everything from the drawing office to Experimental Shop, Racing Dept. and Manufacture. By the way, this factory is quite different from the large-scale plants built by the growing European mass-production firms; it is a comparatively small group of very modern buildings, and houses all the departments close together. Staff and workers total approximately 600, and monthly production oscillates between 280 and 300 cars. In order to get a complete picture of this remarkable organization which is governed by Ferry Porsche, son of the famous Professor, it is necessary to underline that its activities extend to prototype-development for other firms and that in the last months a cross-country vehicle has been developed which embodies many well-proven Porsche features.

The Spyder type 550

One of the main objectives of my visit was to get acquainted with the Racing Sports Porsche "Spyder" type 550. Accompanied by Mr. Klauser, I was able to sample this car both as passenger and driver. First, Klauser demonstrated it to me on the Solitude-circuit, when I was immediately struck by the apparent ease with which fast corners were negotiated, the absolute stability of the astonishingly powerful brakes and the tractability of the 4ohc, 1.5-litre engine. All these points were confirmed when I took the wheel.

I was dropped into a nicely shaped bucket-type seat which was just sufficient to receive my contours, and provided first rate contact with the car even on very high lateral *g*. The left foot rests on a rigid footrest so that it is possible to brace oneself (in Mercedes GP car-fashion with the legs at a rather wide angle) right into the structure of the car. Just behind the steering wheel (which was a very nice Italian affair on the test car) are the instruments, of which a large rev-counter reading up to 10,000 rpm is prominent. Apart from the two electrical fuel pumps (housed to the left inside the cockpit) being switched on by a separate switch, and from the pull-out switches for the two independent ignition systems, the controls are normal. Upon the push of the starter button, the engine came to life with a deep roar, idling at about 1500 rpm.

Starting called for nothing unusual, since the engine doesn't need to be revved up to abnormal heights before taking the car smoothly away, and I quickly changed up and was rounding my first corners. First sensations: The car handles without any trace of either under- or oversteer. I should say that its behaviour is very nearly neutral, which is rather remarkable for a car with swing axle IRS of basically VW geometry. Handling is as if the center of gravity were underground, and there is, of course, no roll discernible while driving. The engine of the "550" must be attributed a double personality: it combines the virtues of a race-bred power package with the character of a reasonably calm sports unit. For driving in racing fashion, 7500 rpm will be used as change-up point (maximum torque happens at 5200 rpm!): going through the gears like this conveys really startling acceleration. On the other hand, the engine can be accelerated from 2500 rpm onwards without difficulty, so that the "550" could and can be used in exactly the same every-day-driving as it is the lot of most Coupé-Porsches, MGs, Austin-Healeys and so on. Only the most normal degree of throttle delicacy is called for.

The brakes are absolutely sensational: It is easy to frighten almost any passenger unfamiliar with this car by approaching a corner at full throttle and, apparently, braking only when it is much too late. There is no reaction on the steering wheel, and the car smoothly slows without having to be checked, while the passenger swallows his stomach . . . The cornering qualities match those of engine and brakes perfectly, and corners of every description can be taken at astonishingly high speeds, without anything special being required from the driver. It is, perhaps, even a bit too easy to go fast with this car, and beginners might do well to stay in constant contact with the speedometer—even so, they will sometimes be alarmed when they see how far the needle has moved. Power-sliding is necessary only when the potential of the car is really explored, but then can be executed with a high degree of accuracy and finesse. Up to very high speeds, there is always surplus power on hand to take the back 'round, and I feel that practicing on various circuits with the "550" is bound to improve the technique of almost any sports car driver below the Fangio-Moss level. By the way, the car I had was equipped with a small racing screen on the driver's side only; while the driver thus enjoys practically the same degree of protection as with a full-width screen, the passenger's head is nearly blown away as soon as 100 mph is exceeded. If you want to remain on speaking terms with your wife, it would be wise to order the "550" with the alternative large screen before taking her out.

About the performance: Although I was not able to execute a full test, I could check the speedometer and then do some performance testing (with two persons aboard, the passenger being exposed in the air stream). For the record: The car had chassis number 550-0048, engine number P 900-54, the engine was equipped with two twin-choke Weber carburetors type 40 DCM 4, the turns of the steering wheel were found to be 2 1/3 from lock to lock. As a works car, the front brakes were 2.36" wide (standard width is 1.57")—with 11" diameter; rear brakes were standard width. Standing start acceleration was measured to be: 0 - 50 mph, 6.2 secs; 0 - 75 mph, 11.3 secs; 0 - 100 mph, 19.2 secs. It must here be pointed out that a) the car would have been quicker off the mark with the standard gear ratios; the car in question was fitted with a non standard "high" first gear which made getting off at full bore not too easy, b) that at higher speeds the

At left, the instrument panel includes a 10,000 rpm tach and two pull-out switches to control the double ignition. Adjacent picture shows the new 3-piece crankcase with Hirth crankshaft. Below left, the upper half of the crankcase has been fitted and bolts are torque-tightened, while below a Super engine is installed in a Coupe.

unprotected passenger noticeably affects the performance. Top speed could not be tested, but I have no reason to doubt the factory's statement that this lies between 135 and 145 mph. Weight is 1300 lbs dry, which gives a power to weight ratio of 186 bhp/ton.

Chassis and engine

The layout of the 550 is derived from that of the series-production type 356. In place of the chassisless-construction, a classic rectangular frame built up from steel tubing has been adopted. Front suspension is by Porsche trailing links combined with an anti-roll torsion bar. In the rear, the most noticeable change is that the relative position of engine and gearbox is reversed in comparison to the coupe: The engine lies in front of the rear axle, behind which comes the gearbox. The weight of the most heavy items is thus concentrated towards the center of the chassis, so that a uniform weight distribution of 50/50 is achieved; the polar moment of inertia is low which makes for a high degree of responsiveness. The suspension of the rear wheels is basically the same as that on the standard Porsches, with the geometry altered so that there is less camber (with the car fueled up and the driver in it, there is even some negative camber) than on the production types. This is the same expedient that has been adopted by every swing-axle producing European sports- or racing car manufacturer to improve the handling qualities; the first was Alfa Romeo on their GP car type 158 (the "Alfetta").

The engine is a very remarkable piece of engineering and certainly the most powerful air-cooled power unit of only 1500 cc capacity. Its design is based on that of the standard engines with a light-alloy crankcase split on a vertical plane. The built-up Hirth-crankshaft runs on three twin-row roller bearings and carries con-rods with one-piece twin-row roller bearing big ends. Below and parallel to the crankshaft (where the standard engine has its camshaft) lies a sub-shaft driven by a pair of spur gears; at the nose of this shaft two pinions mounted side by side drive two transverse shafts that run, parallel to the cylinder axis, outwards just below the cylinder barrels. Each transverse shaft drives, by means of a further pair of pinions, one of the exhaust camshafts. At the other end of the exhaust camshafts, further pairs of pinions convey the drive to near-vertical shafts that drive the (upper) inlet camshafts. To sum up: There is 1 sub-shaft parallel to the crankshaft, 2 transverse-shafts parallel to the cylinder axis, 2 near-vertical shafts taking the drive up to the inlet camshafts, and, of course, 4 overhead camshafts. Complicated, perhaps, but a) it is mechanically simple, b) it allows for the rather great heat expansion of the air-cooled light-alloy cylinders, c) it enables a crankcase to be used which is practically identical with that of the standard-production ohv types.

Detail Work

The engine of the "550" contains many unusual and individual design details. Naturally, all the visible and discernible features do not answer the famous "what-makes-it-tick" question 100 per cent, because experience is very important, particularly in the case of an air-cooled unit with a specific output of 74 bhp per litre—this was equalled by Ferrari (I am speaking of engines running on pump fuel, remember!) in 1953 only, when it was thought to be outstanding—or about the same as that of the direct injection.

Among the readily discernible items: Cooling-air blower with fan having two rows of fins; special twin-row roller bearings for crankshaft and big ends; aluminum-alloy cylinder barrels (four individual) with chromium-plated bore as on the production-types; light-alloy cylinder heads with hemispherical combustion chambers; two valves per cylinder and two spark plugs; two ignition coils and two individual distributors driven by the inlet-camshafts; valves actuated through fingers (these present an interesting design detail since they are not mounted on a common shaft but pressed down onto individual bolts with hardened spherical bearing-heads by means of springs —an idea similar to the rockers on the V8 Chevrolet). For those who can afford it, the "Spyder" represents a design which leaves very little scope for improvements. ●

Above: every Porsche steering gear is automatically "run-in" for 10 hrs. Below, the entire rear body section is hinged and removable.

Cross section of the racing engine shows the roller type connecting rods and the overhead cams operating the valves via followers held on pring-loaded bearings.

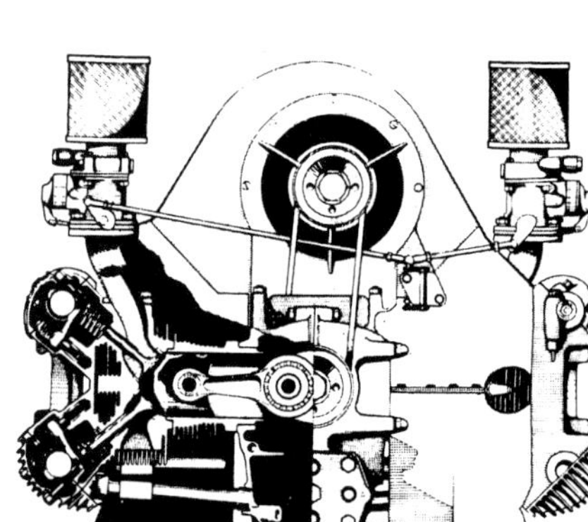

Rear view of the Spyder: prototype's fender fins have nearly vanished.

South German Test-outing

CRESTING a rise on the Autobahn spur which leads to Heilbronn, the Porsche Spyder competition 2-seater for which over 135 m.p.h. is claimed has headlamps blazing to warn other traffic of its approach.

A TRIO OF PORSCHES SAMPLED

By Joseph Lowrey, B.Sc.(Eng.)

BETWEEN the B.M.W. and Porsche companies there is at least one thing in common, the fact that both started virtually from scratch after the war. Whereas the B.M.W. factory at Munich is setting out to serve a really big market with the Isetta coupé, however, the Porsche factory in a Stuttgart suburb is dedicated to the much more specialist task of producing sports cars.

From its inception in modest premises where Volkswagen components were built into a sports coupé of finely streamlined shape, however, the Porsche company has advanced a very long way. In the cars, ideas inherited from the late Ferdinand Porsche's design for the Volkswagen are still evident, but of actual VW components there are now virtually none. Likewise, the original factory has now given place to a new one close at hand which, while still moderate in size, makes very bold use of a hillside site to employ eye-catching yet practical unorthodoxies of architecture. Porsche sell more than half their production to Americans, many to those stationed in Germany, who can therefore visit the factory.

Work as consulting engineers on various projects is still continued actively by the Porsche company, a prototype military vehicle with air-cooled rear engine and four-wheel drive having already been unveiled, and an air-cooled V-type engine being evident in one corner of a works normally associated with flat-four power units. Sports cars are the main business, however, graduated from normal road-going coupés and cabriolets through those similar-looking cars with two more potent types of engine to the "Spyder" 2-seater which is designed primarily for racing.

Production capacity in the Porsche works is quoted as 14 cars per day, although at the time of our visit the recently announced 1956 models were as yet only emerging at about half this rate, due to delays in obtaining castings and other outside components. Quite elaborate tooling for the machining of cylinder blocks, crankcases and other parts was evident, but the policy of having one craftsman responsible for assembly of a particular engine almost from scratch is adhered to, the neat and clean benches being most impressive. Trimmed bodies (this is a chassisless steel car) from the Reutter factory next door are mounted on high trolleys, and moved in line through the works as mechanical components are assembled on to their underside, each car then undergoing about 60 miles of testing on the nearby Autobahn before delivery. "Trade plates" mounted on an apron which covers the whole front of the car guard against damage to paintwork during these tests.

Most buyers of Porsche cars are urged strongly by the makers to choose the normal engine, which uses plain big-end bearings. The rather similar looking Super engine with roller big-end bearings and higher power output offers more performance for use in competitions, but has

NORMAL version of the latest 1.6-litre coupé, the Porsche seen here proved itself a 100 m.p.h. car in which the low level of wind noise was especially notable.

a shorter life between overhauls and is noisier and less flexible, disadvantages which are thought to outweigh extra speed for most buyers. The Carrera model, with a slightly modified version of the "Spyder" four-camshaft engine is even more specialized.

Externally the Porsche changes comparatively little from year to year, care being taken to avoid spoiling a good basic shape. Internally, it has altered a good deal for 1956, perhaps the most valuable change being an enlargement of the cylinder bore to step up the displacement from 1,488 c.c. to 1,582 c.c. Also, the torsion bar suspension has been softened quite considerably and provision made for longer wheel travel, softer springing increasing the effectiveness of the front-end anti-roll torsion bar which was introduced last year, and being matched by improved shock absorbers.

With a revised model not yet in full production, up-to-date cars were scarce around the factory, but after a drive in the 1.5-litre car which Klaus Rucker (a Porsche engineer lately returned to Germany after a period with Studebaker in America) was using for personal transport, we were chased away by von Hanstein for a spell of motoring in a 1.6-litre car, which had not yet finished routine testing before the carpets were installed and it was delivered to a dealer.

Having quite a personal liking for cars which can be "wished" around corners without appreciable physical "steering," my first reaction was to prefer the handling of the 1955 car to the slightly more "Americanized" controls of the 1956 model. After a wild dice through the back-streets of Stuttgart in the latter, however, with von Hanstein in the passenger seat urging me to greater efforts in the hope that we might get ahead of Uhlenhaut who was hustling more conventionally down the busier main road in his Mercedes 300 SLR coupé (we failed by a narrow margin) I soon forgot my reactionary ideas.

By the standards expected in England, the Porsche is by no means flexible in its performance. On the rev. counter there is marked a green sector from 2,500 r.p.m. to 4,500 r.p.m., a range of "desired" speeds which means from 13 to 24 m.p.h. in 1st gear, from 24 to 43 m.p.h. in 2nd gear, from 37 m.p.h. to 67 m.p.h. in 3rd gear, and from 51 to 93 m.p.h. in top gear. There is not much overlap in these speed ranges, to give a driver freedom of

FASTEST product of the Porsche factory, the Spyder 2-seater has its engine ahead of the rear axle instead of behind it, and uses a multiple-tube chassis frame whereas touring models have combined steel bodies and chassis.

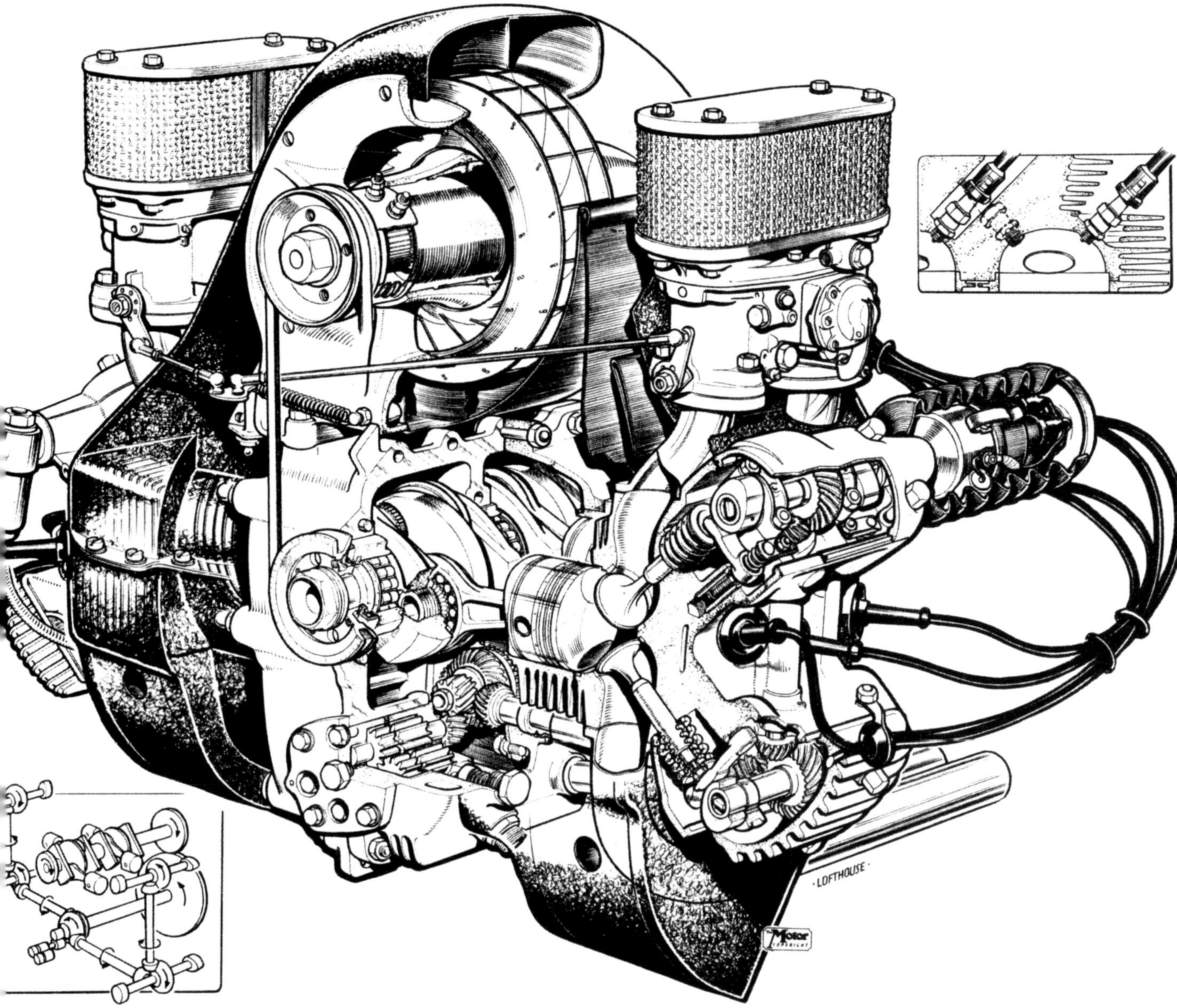

AVAILABLE with small modifications in normally bodied cars, the Spyder competition engine is shown in this drawing by an artist of *The Motor*. Air cooling of four horizontally opposed cylinders is retained, but pushrod valve operation gives place to four overhead camshafts driven by shafts and bevel gears. Roller main and big-end bearings are used on the Hirth built-up crankshaft, carburation is by a pair of twin-choke downdraught carburetters, and two plugs per cylinder are sparked from distributors on the rear ends of the two inlet camshafts.

choice about when to change gear upwards or downwards. In actual fact, however, with the larger and more flexible 1956 engine the speed can drop as low as 1,800 r.p.m. before snatch becomes evident, and the red sector on the rev. counter from 4,500 r.p.m. to 5,000 r.p.m. seems to be intended merely as a suggestion that sustained cruising at these speeds is not advisable.

My impression, incidentally, was that it was not really the engine which was in any way inflexible, but that the use of a short and "solid" transmission line was probably making the use of low r.p.m. a jerky process. A spring-centre clutch, long propeller shaft and axle mounted on "cart" springs together provide flexibility which if cleverly used can reduce the "resonant" speed of a transmission to a very low figure.

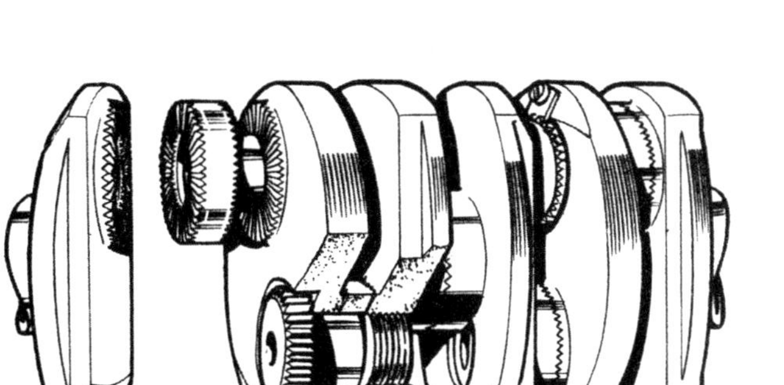

South German Test-outing - Contd.

STARK simplicity characterized the cockpit of the sports-racing Spyder which was sampled, although some cars are built with touring equipment for use in Rallies. Location of the engine immediately behind the driver, with the spare wheel enclosed above the gearbox, will be noted.

What makes a relatively narrow band of really useful engine speeds perfectly acceptable is, of course, the fact that there is no undue fuss at the higher r.p.m., and that the gearbox is a delight to use. The brand-new engine of the car we were driving would not mind short periods of hard driving, we were assured, so we decided to take a few performance figures, two-up but without our usual weighty box of test apparatus, along the least-hilly piece of the Stuttgart to Heilbronn Autobahn which we could find. Checking the kilometres-per-hour speedometer at speeds between 30 and 80 m.p.h. equivalent, we found fairly consistent optimism of 6% or 7%, so having found an indicated 170 k.p.h. to be attainable either way on the level decided that this (non-Super) model was indeed just about a true 100 m.p.h. car as claimed.

Against the Watch

So far as acceleration went, the 1,600 c.c. Porsche had very nearly constant urge through its modest range of useful speeds. Space between road undulations was limited, but the two-way average time from 30 to 50 m.p.h. was 15.4 sec. in top gear or 8.5 sec. in 3rd gear, the corresponding times from 40 to 60 m.p.h. were 16.6 sec. and 9.1 sec. respectively. From a standing start, using an indicated 5,000 r.p.m. in each gear, the timed acceleration figures were: 0-30 m.p.h. in 5.1 sec., 0-40 m.p.h. in 7.7 sec., 0-50 m.p.h. in 11.1 sec., 0-60 m.p.h. in 16.6 sec., and 0-70 m.p.h. in 25.3 sec., with a standing quarter mile covered in 19 sec. Balancing the newness of the car against its having slightly less weight aboard than it would carry if subjected to a proper Road Test Report, these are commendably good figures, especially the standing ¼ mile time which benefits more than the others from a no-wheel-spin getaway with plenty of weight on independently sprung rear wheels.

Whilst the Porsche is not exactly silent mechanically, it is most notably free from wind noise when running quickly on the Autobahn, and the new springing gives a comfortable ride without introducing perceptible roll to the cornering characteristics. There *is* a back seat, and to record acceleration times I sat in (or more precisely, across) it, after the driving seat had been slid well forward, but I was very glad to escape from that cramped place at the first opportunity. Call it a two-seater, with room at the back for small children or a lot of luggage, nicely furnished as a businessman's rushabout.

Even Faster

One more outing was provided for us before we left Stuttgart, an outing which contrasted strongly with the comfortably brisk travel in the Porsche 356A/1,600 coupé. One of the four-camshaft Spyder racing 2-seaters, with its engine mounted within the wheelbase and carrying very simple full-width bodywork on its tubular chassis, was held back for a spell from being stripped for a change of axle ratio, and we were each given a ride up and down the Autobahn. Traffic did not allow the speedometer needle to go round to the claimed top speed of 220 k.p.h. (137 m.p.h., on 1½ litres), and without the shelter of even an aero screen in front of the passenger seat the tasks of keeping on breathing and of avoiding being blown into a reclining position distracted us from any attempts at very scientific observation. But, if the speedometer was to be believed, the great surge of acceleration from rest had not quite faded out as 200 k.p.h. (124 m.p.h.) was exceeded, the works test driver complaining that the bulk of a passenger was slowing the car considerably. With this same engine, slightly decompressioned and de-tuned but quoted as giving only 10% less power, installed in one of the sleek hard-top cars, remarkably impressive performance should be provided. although traffic might be quite something of a headache,

Unexpectedly, we found that our return journey coincided with a one-day public holiday in Germany, which frustrated a few odds and ends of shopping but provided compensations by greatly reducing the number of heavy lorries on the Autobahn. After a detour on to ordinary roads across the fog-filled Rhine valley to Bonn for lunch, and a teatime stop at Arnhem, we reached the Hook of Holland in comfortable time to eat a very adequate meal in the station restaurant, then embarked early for the return to an England which, while we had been "enjoying" alternate fog and drizzle, had unkindly staged in our absence some of the kindest weather of the winter!

HEAD MAN of Porsche is Dr. Ing. Ferry Porsche, son of the famed Auto Union and Volkswagen designer, Dr. Ferdinand Porsche.

COMPETITION DEPARTMENT of Porsche is in the main office block, a building only recently de-requisitioned by the U.S. Army. In the foreground are two of the "works" team of Type 550 1,500 c.c. Spyders. They put up a remarkable show in last Sunday's Nürburgring 1,000 kms. race.

CHEZ PORSCHE

On a Recent Visit to Stuttgart, Maxwell Boyd took these Photographs in the Competition Department of the Porsche Factory

COMPETITION MANAGER, and highly successful Porsche driver, Baron H. van Hanstein, conferring with some of his mechanics about a Spyder being prepared for a forthcoming long-distance race.

FINELY MACHINED is this Spyder brake drum, with each fin drilled for cooling and weight-saving. The dark patch between the fins is a balance weight.

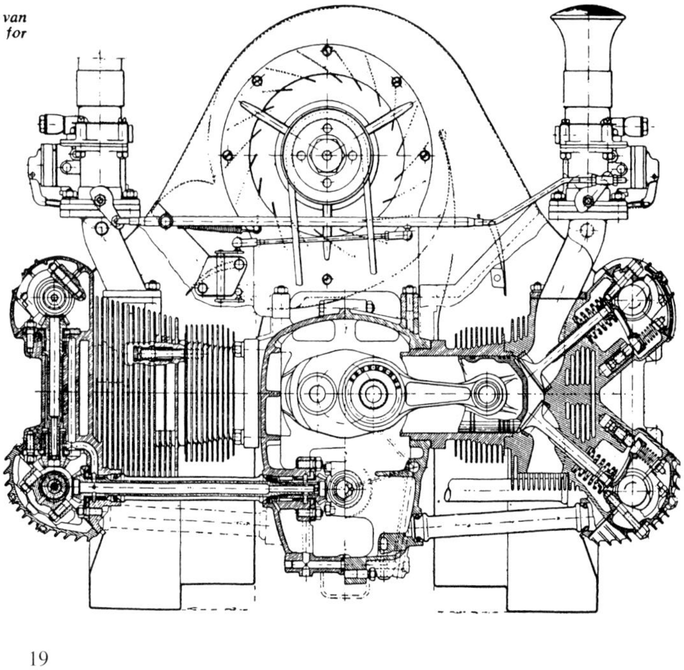

FUEL TANK (above) of the Spyder is mounted in front of the driver and has a "cutaway" to accommodate the spare wheel.

HIGHLY SUCCESSFUL (right) in racing all over the world is the 4-cyl., 4-o.h.c., 1,500 c.c. Type 550 engine, which is also fitted to the Carrera model.

Owner Porter puts his R.R. cap on backwards for the high speed runs.

The aero-screen is small but effective in keeping air-blast off the driver.

Road Test: PORSCHE 550 SPYDER

"CHANGE is progress," as the saying goes, and this axiom applies particularly well to the field of competition sports cars. Introduced a little over two years ago, the Porsche Spyder type 550/1500 RS immediately proved itself to be a formidable competitor in the under 1500 category, and the Stuttgart factory soon made the car available to enthusiasts the world over.

To date, over 100 of these beautifully engineered machines have been built and sold, and it is pertinent to note that despite the oft-mentioned "edge" of factory-prepared machines, the 550 Spyder has completely dominated its class, both here and abroad, even in the hands of private owners.

Although we could have tested one of the ex-factory team cars, we felt that the performance data of an absolutely stock, privately owned machine would be a more interesting, more honest test, truly indicative of what kind of performance can be purchased "off the shelf." Many of the Spyders now winning the big races are slightly modified and obviously out-perform the normal 550, but at least you can buy a car such as this. However, a new competition model to replace the 550 is rumored to be in the offing.

Our test car was generously supplied (and capably driven) by its owner-enthusiast, John H. Porter, Jr. of Newport Beach, California. This machine has run in practically every California event offered during the past two years and the odometer read 8400 miles, when we began the test. Mechanically, it is exactly as purchased with two exceptions. The brakes are the latest (larger) type and, as tested, it was set-up with the short-track gear ratios just as it came off the course at Palm Springs. The axle gears were 34/7 (4.855), giving a final drive ratio in 4th gear of 4.66 to one in place of the standard catalogued ratio of 3.56. In effect this alteration gives substantially better acceleration but limits the ultimate top speed capability to the driver's decision as to what constitutes a safe engine revolution rate.

Accordingly, we made only one high-speed run, and with the tachometer at nearly 7500 rpm the car recorded 121.6 mph. The factory states that the car is capable of 140 mph and there is no reason to doubt this figure—given the "high-speed" rear axle ratio. The car was tested with about 10 gallons of normal premium fuel in the tank and no fuel consumption checks were made. The owner drives the car occasionally on the highway and reports up to 16 mpg with as low as 10 mpg under the stress of competition. The curb weight given (1510 lbs.) was a little surprising but was taken with a full fuel tank (24 gallons). Weight with no fuel would be 1366 lbs. or slightly less than the catalogued "empty" weight given as 1407 lbs.

Aside from the truly remarkable acceleration figures given in both tabular and graphic forms, the most interesting feature of the 550 is, without a doubt, the engine. This unit was described in R & T as far back as May, 1954 and further described in January, 1956. Suffice it to say that the combination of a very short stroke, the Hirth roller bearing crankshaft, and no less than four overhead camshafts (two for each cylinder bank) makes for a very free-running power plant and one which never seems to be approaching

The weighing-in ceremonies on State certified scales. 1510 lbs. with full tank.

Porter explains to the editor that he has no trouble with retreads.

Despite all the machinery being at the rear the weight distribution is 48/52.

Top contender in the 1500cc class

the bursting point, even if 7500 rpm is exceeded occasionally. This speed was in fact used as a rev limit and is equivalent to a piston speed of only 3245 fpm.

In addition to the engine's phenomenal ability to rev smoothly and safely at over 7500 rpm, it should be mentioned that the unit is remarkably tractable for use off the race circuit. True, the tachometer is plainly marked so as to indicate that the engine should be operated between 4000 and 7500 rpm—no more, no less. Actually the unit idles at 1200 rpm and pulls steadily in any gear from this point. However, when accelerating, the engine feels rather tame until 3500 rpm is reached at which point the Tapley meter advances rapidly to a much higher reading and continues to increase slowly until approximately 5500 rpm (the torque peaking speed) is attained.

Driving the 550 is very little different from handling a stock production Porsche Speedster or coupe. The steering is the same (2.3 turns, lock to lock), and the control of clutch, brakes, and gear lever will feel familiar to any Porsche owner. Furthermore, the seats are very similar to those supplied with the Speedsters and are quite comfortable except for the obvious fact that the passenger is rather exposed to the elements. Of course the performance literally forces you back in the seat, but surprisingly the thrill wears off rather quickly, especially on a lonely road with little or no traffic. In congested areas one quickly learns that this car gets above the legal speed limit so easily in 2nd gear that one is very likely to misjudge the speed. The first time this happens the brakes come into their own, for they are among the best we have ever tried. Applied hard at over 100 mph they bring the speed down so rapidly that the passenger stays seated only with difficulty. Competition experience with the 550 has proven that these brakes are virtually trouble-free and among the best when it comes to fade immunity.

The steering characteristic is near-neutral with the usual high performance car feature of a rear-end breakaway controlable by the throttle foot. Both the above and the 48/52 % fore-and-aft weight distribution were somewhat surprising to us, as we expected some oversteer and a weight distribution closer to 45/55. However, as is well known, the 550 engine lies ahead of the rear axle, the gearbox behind—the exact reverse of a "normal" Porsche. This new twist and the initial negative camber of the rear wheels accounts for the good handling of the 550. We did not drive the car enough to make any evaluation of the low polar moment, but it would appear that the more central concentration of masses does make for a machine which responds very quickly to the driver's merest whim—the old idea that you can almost literally "wish" a good-handling car through a fast curve.

Combining as it does, terrific performance, faultless handling and excellent brakes, it is no wonder that the Porsche Spyder type 550/1500 RS is the car to beat these days, in its class. ●

R & T ROAD TEST NO. 123

PORSCHE 550 SPYDER

SPECIFICATIONS

List price	$6800
Wheelbase, in.	82.7
Tread, f/r	50.8/49.2
Tire size, f/r	5.00/5.25-16
Curb weight, lbs.	1510
distribution, %	48/52
Test weight	1770
Engine	flat 4, dohc
Bore & stroke	3.35 x 2.60
Displacement, cu in.	91.4
cu cm.	1498
Compression ratio	9.50
Horsepower	137
peaking speed	6200
equivalent mph	103
Torque, ft-lbs.	95.5
peaking speed	5500
equivalent mph	91.4
Gear ratios, overall	
4th	4.66
3rd	5.95
2nd	8.56
1st	15.4

CALCULATED DATA

Lbs/hp (test wt.)	12.9
Cu. ft/ton mile	108
Engine revs/mile	3610
Piston travel, ft/mi.	1560
Mph @ 2500 fpm	96.1

PERFORMANCE, Mph

Top speed, one-way	121.6
3rd (7500)	97.5
2nd (7500)	68
1st (7500)	38
see chart for shift points	
Mileage range	10/16 mpg

ACCELERATION, Secs.

0-30 mph	2.6
0-40 mph	4.6
0-50 mph	6.2
0-60 mph	8.2
0-70 mph	11.6
0-80 mph	15.0
0-90 mph	19.6
0-100 mph	25.2
standing start ¼ mile	16.1

TAPLEY DATA, Lbs/ton

4th	220 @ 85 mph
3rd	360 @ 67 mph
2nd	540 @ 50 mph
1st	off-scale

Total drag at 60 mph, 69 lbs.

SPEEDO ERROR

Indicated	Actual
30 mph	30.0
40 mph	39.6
50 mph	48.7
60 mph	57.5
80 mph	75.6
100 mph	95.8
123 mph	121.6

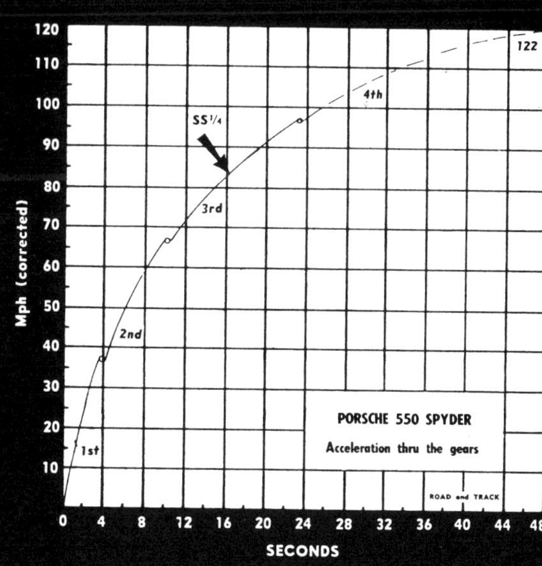

SCI Technical Report:

The Fastest PORSCHE

By GRIFF BORGESON with KEN MILES & RICHIE GINTHER

Photos by Don Typond

THE Porsche Spyder is a fantastic machine. It is fast and fiercely powerful, one of the most successful competition sports cars in history, and it has a personality no less complex than Dr. Jekyll's. At brisk highway speeds it is very easy to drive, docile, perfectly manageable, superbly responsive. It is even an easy car to drive "fairly" fast, the way Ken Miles and Richie Ginther have driven it in a long series of spectacular 1500-cc-modified wins. But to drive it really fast, as fast as it's able to go—this is work for a driving genius—for a man, as Miles puts it, "with extraordinarily quick reactions and an exceedingly delicate bottom." Says Miles: "I'm just not that much of a driver."

At the same time, there is nothing in the car's character that makes it inappropriate for use as a round-town hack. It has enough street equipment to satisfy the least austere sports car fan: top, windshield, trafficators, speedometer, windshield wipers, everything but R, H and OD, as they say in the used-car ads, and there's nothing to keep you from having radio and heater installed. Miles has a Spyder that he uses daily in L. A. traffic, and he says, "Why not? Set up for town driving, with soft plugs and the right jets, it's a joy to drive—comfortable, roomy, and it goes like a bullet from a gun when you open the tap."

I can vouch for part of his description—the last part. But SCI's test car, Richie Ginther's hurtling number 211, was hardly in street tune. The hard plugs were still in it from Palm Springs. The ¼ inch bored out intake ports and the whopping dual-throat Webers, mounted in place of the standard Solex carbs, reduced low-rpm fuel-air velocity to a near minimum. In the interest of better weight distribution, Ginther had gutted the cockpit and moved the spare tire and wheel in from their normal position aft of the engine. Sharing the passenger side with these impedimenta made the word "roomy" no longer accurate. With the big red numbers on the side, Ginther's Spyder didn't even *look* like a street machine. And there was, of course, the exhaust.

Richie Ginther bends the Spyder through some tight turns on the test course used by SCI.

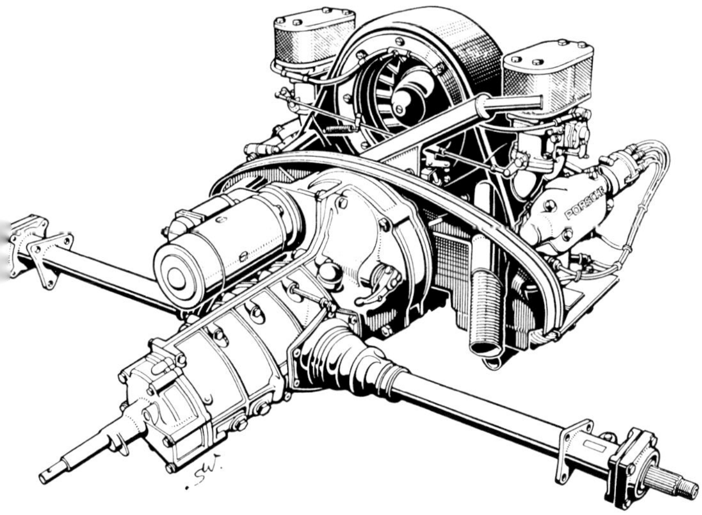

Spyder engine coupled to transmission and final drive unit. Dual throat Webers used on Ginther's car are optional in place of Solexes.

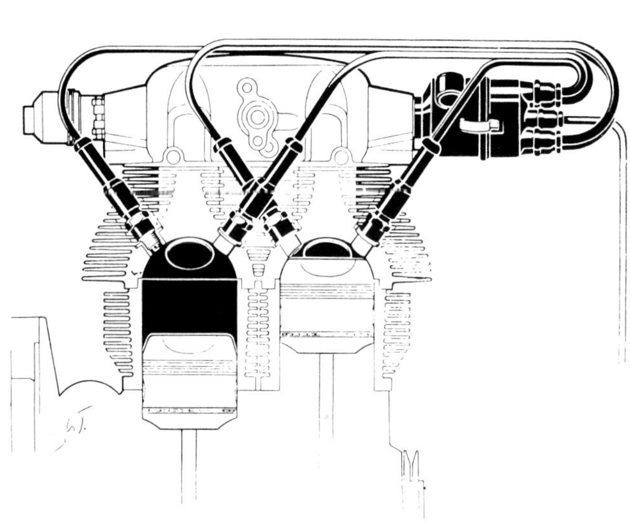

Ignition layout. Two spark plugs fired simultaneously eliminate chances of misfire and ignite fuel from both ends of chamber.

In stock form the Spyder is fitted with expansion chambers in the forked exhaust pipes and a perforated can in the single tailpipe, and it's a reasonably quiet machine. But with a set of large diameter, straight-through pipes the sound is shattering. A single 4000-rpm blip on the throttle and you expect to see cracks spreading in nearby concrete walls. The exhaust note starts as a resonant, thundering hammer and builds to that indescribable ripping-canvas scream that can come only from a barrel chested 7500 rpm.

It was in this inconspicuous fashion that Ginther and I took off through Hollywood in heavy morning traffic, in an aura of fragrant Castrol fumes. To reach the coast in order to bear north we had to pass through Beverly Hills, a town that takes pride in the strict enforcement of its anti-noise laws. It's a handsome tribute to car and driver that we ran the gauntlet without accident. The Spyder engine stroked along docilely at 1500 rpm in top gear! In spite of its steep power and torque curves it felt enormously strong at low revs, strong enough to make other 1500 cc road machines seem feeble by comparison.

Once we hit the back country, though, we could give the beast its head. The first maneuver was to take it through the first three cogs up to 7500 rpm, to clean out the residues of an hour's poking through traffic. The savage, lunging character of the car under full throttle is unforgettable. Up to about 5000 rpm it feels like one of the thrustiest machines you've ever driven, but then the cams hit their stride and the power *really* comes on. The most useful power range is concentrated in a narrow band between about 6000 and 7000 rpm.

"It's hard," said Ginther, "for a guy getting out of a pushrod Porsche to learn to keep the revs up. In a normal Porsche you get up to 5500 and the valves bounce. A push-rod man in a Spyder invariably feels that he's starting to wind the engine pretty tight long before he gets there. This car is geared for about 125 mph at 7500, and I turn to

TEST CONDITIONS:

Number aboard	Driver only (during speed runs).
Top position	Removed
Temperature	[illegible]

TOP SPEED:

Two-way average	128.3
Fastest one-way run	129.6

ACCELERATION:

From zero to	Seconds
30 mph	3.4
40 mph	4.6
50 mph	5.2
60 mph	7.2
70 mph	9.5
80 mph	13.2
90 mph	15.0
100 mph	19.5

PULLING POWER: (Tapley Meter)

	Lbs/Ton	Grade %	Mph
I	off scale	..	..
II	600	31	41
III	460	24	58
IV	250	13	75

Drag at 10 mph	10
Drag at 60 mph	55

SPEEDOMETER CORRECTION:

Indicated	Actual
30	28.8
40	36.5
50	45.0
60	53.2
70	61.6
80	70.3
90	78.2
100	87.0

FUEL CONSUMPTION:

Hard driving	18.5 mpg during entire test.

BRAKING EFFICIENCY: Summation – zero fade.

SPECIFICATIONS

POWER UNIT:

Type	Four-cylinder opposed, air-cooled.
Valve Arrangement	Vee-inclined valves, d.o.h.c.
Bore & Stroke (Engl. & Met.)	3.35 x 2.60 ins; 85 x 66 mm.

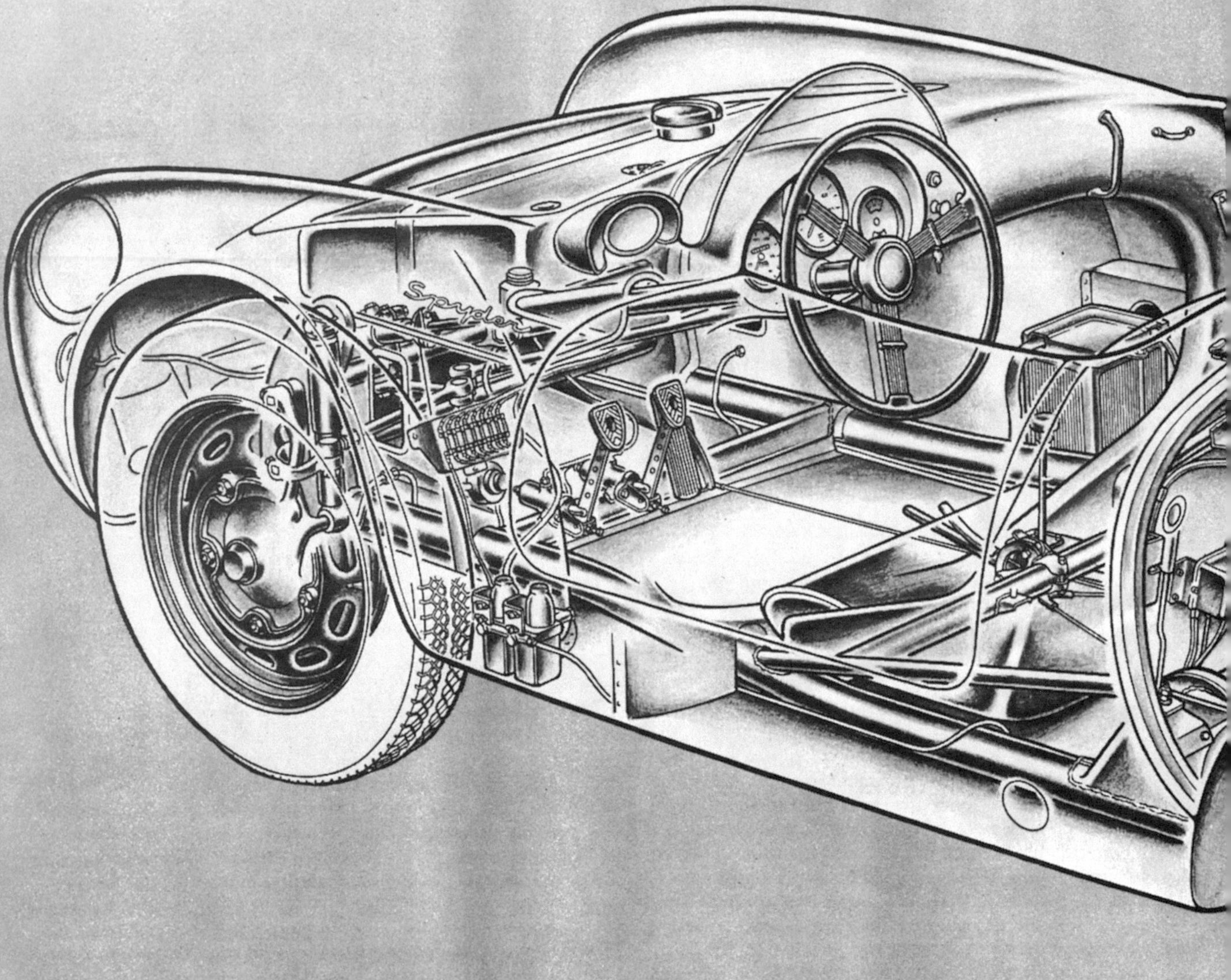

PORSCHE SPYDER "... like a bullet when you open the tap."

Stroke/Bore Ratio	0.78 to one.
Displacement (Engl. & Met.)	91.4 cu. ins; 1498 cc.
Compression Ratio	12.0 to one.
Carburetion by	Two dual-throat Webers.
Max. bhp @ rpm	130 @ 7000 (approx.).
Idle Speed	1500 rpm.

CHASSIS:

Wheelbase	82.7 ins.
Front Tread	51.4 ins.
Rear Tread	50.0 ins.
Suspension, front	Transverse torsion bars (laminated), fully adjustable.
Suspension, rear	Transverse torsion bars (solid), fully adjustable.
Shock absorbers	Telescopic double acting; one per wheel.
Steering type	ZF-Gemmer Worm and Ball.
Brake type	ATE hydraulic; two leading shoes front.
Brake lining area	165 sq. ins.
Tire size	F500 x 16; R525 x 16.

GENERAL:

Length	141.75 ins.
Width	61.0 ins.
Height	40.0 ins.
Weight, test car	1480 lbs. (full fuel tank).
Weight distribution, F/R	48.6/51.4
Weight distribution, F/R, with driver	49.1/50.9
Fuel capacity	17.2 U.S. gallons.

RATING FACTORS:

Bhp per cu. in.	1.42
Bhp per sq. in. piston area	3.69
Pounds per bhp, test car	11.4
Piston speed @ 60 mph	1564 fpm.
Piston speed @ max bhp	3033 fpm.
Brake lining area per ton, test car	223 sq. ins.

TRANSMISSION DATA

Gear	Trans. Ratio	Tooth Comb.	Overall Ratio	Mph per 1000 Rpm	Mph @ Max bhp	Max Test Speed	Equals Rpm	Equals Ft./Min.
I	2.54	13/33	12.34	6.3	44	47	7500	3250
II	1.76	17/30	8.55	9.1	64	68	7500	3250
III	1.23	22/27	5.98	13.0	91	97	7500	3250
IV	0.96	25/24	4.67	16.6	116	129.6	7800	3380

Final drive ratio, test car	4.86; 7/34
Other available final drive ratio	4.43; 7/31
Axle torque taken by	Radius arms.

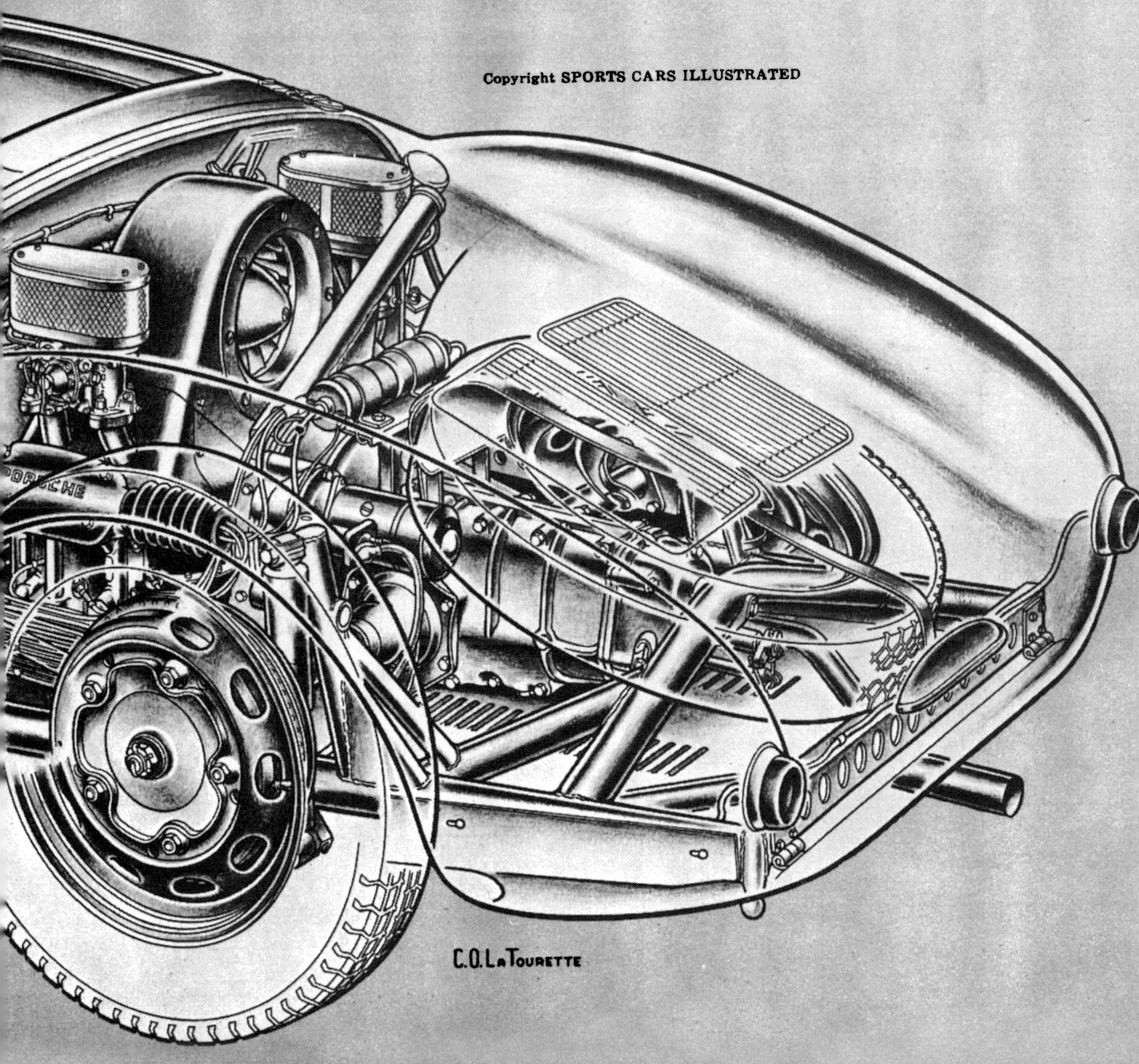

Richie Ginther gives Stu Hilborn the lowdown on the big downdraft Webers. Stressed sheet housings over rear shocks supplement original big-tube supports.

Though it's strictly for go, as shown by the drilled pedal arms and exposed frame tubes, the Spyder has a road-car finish This one has directional signals!

ABOVE: Twin plugs are angled and widely spaced, ensure complete combustion. They're placed near valves and seats, to minimize local hot-spotting. Ports are big, but more important, their sectional area is constant throughout. Shaft drive to cams can be seen at top.

BELOW: Deep bevels in healthy piston crowns provide clearance for misbehaving valves. Cylinder finning is symmetrical as possible to prevent uneven expansion of bore. Pump and connections for dry-sump oil system are at lower right. Cam drive leaves sump at center.

7800 regularly. At Palm Springs I tried to keep it above 5500 coming out of all the corners. There's a 4200 ft. straight-away on the course, and if I came out onto it clean and fast at about 80, I turned a good 8000 rpm before I had to shut off. We know we can turn 8200 without valve gear trouble, and even at this figure piston speed is no problem."

Learning to drive by the tach and ignore the scream of the d.o. engine may be tough, but it's childishly simple compared with taking the car around a race course at something approaching its limits of adhesion.

"The Spyder," Miles says, "isn't like, say, a Maserati 1500 —a car that anyone can get into and drive right on the ragged edge. With an ordinary car, when you're cornering fast there's a point where the car is perfectly stable, a point where the car is gone, and a substantial transition point, where you know it's starting to go and can do something about it. With the Spyder it happens *now*. A fairly good driver has no trouble taking it up to the point where it's nearly gone—but the brilliant driver, the genius, can add 5 mph to the cornering speed beyond that point. I'm referring here to a Kling or a Moss, not to myself."

You'd never, of course, discover limits on the road. At speed you can't even detect that the Spyder has a swing axle. It rides dead-flat and stays glued during terribly fast cornering. In a 30 mile ride over twisting mountain roads in Ginther's car, I noticed no sign of that usual "Porsche characteristic"—the slippery rear end. There's a lot less change of track in the Spyder than in the ordinary Porsche. Its very considerable negative camber brings the pivot point of the swing axle closer to the road. A line drawn through the point of tire contact with the road is more nearly parallel with the road, so that the contact points move more nearly straight up and down.

You do notice the swing axle on a fast corner with an uneven surface, where you get great side forces combined

Ken Miles feels the Spyder is a fine street machine, and is setting out to prove it. Single screen, twin rear view mirrors and fast-open oil cap reveal reality.

with a lot of wheel movement. This, added to the fact that the front suspension flexes a lot under these circumstances, makes the car a hell of a handful on a fast, bumpy course. The unsprung weight in the rear is fairly high relative to the weight of the car; it includes the big axle tubes, brake drums and radius arms. Among the effects of all the weight is a good deal of rear wheel patter when you're braking hard on rough surfaces. A lot of this has been eliminated in the new space-frame Spyder. The frame is a much stiffer structure, which helps the front suspension, and the rear suspension *always* has negative camber.

Of the Spyder's handling qualities Ginther has this to say: "The fastest way around a slow hairpin turn is a matter of driver opinion. But on any fast turn the straightest, smoothest line is the fastest way. You can put a man in it who can dirt-track all over the course, but a straight-line man like Miles will drive right by him and not even look fast.

"If the Spyder's tail comes loose you can just leave it hanging out. You won't go as fast as when it's behind you, and I know that for sure. But you'll get into no trouble, and you can ignore it—which you'd better not do in the 1600 Speedster. Once the front *and* rear wheels start sliding, though, you're in a different situation. You have to anticipate how much room you have left, how much of it you're going to use, and manipulate throttle and steering accordingly."

The Spyder's steering is as light as they come. The brake pedal and clutch pedal pressure are light—very light. Throttle response is good, particularly when you approach the engine's peak-power operating range. There's nothing to changing gears; you shift with a true flick of the wrist. All the pedals are wide-spaced, and there's no doubt about which foot is on what. The ride is soft. There is no high-

Oil tank is fitted with filler, vent, temperature pickup and connections to engine and front-mounted radiator. It's held by a felt-lined steel strap.

Valve fingers pivot on spherical surfaces, are retained on vertical rod by small spring and bracket at right. Valves, springs, shims, retainer and cotters, and cam are here.

Still housed in the Volkswagen case, the Spyder gearbox offers a wide ratio choice and wrapping-ring synchro. A ZF differential is fitted here, has been no trouble.

frequency noise audible in the cockpit. The screaming duet of engine and exhaust is carried away behind you. The result of these facts is that the Spyder taxes the driver very little. While a one-hour race can leave a Ferrari driver too weak in the legs to walk without limping and too weak in the arms to pick up his trophy, the Spyder leaves him with no more than a reasonable sweat.

For this reason, plus the fact that the Spyder is a heavy, solidly-built machine, it is difficult to contradict Miles when he calls the Spyder the finest long-distance race car in the world.

"A Spyder will go round and round a desperately bumpy course like Monza," he says, "and nothing will fall off. Meanwhile the lightweight racing cars in its class are likely to have broken in half. The Spyder is strong. For any long race—for Mexico, Rheims, Le Mans, Nürburgring—nothing can touch it. On courses like these the low driver-fatigue and even the high but narrow peak-power range pay off.

"The engine, if you don't over-rev it, is entirely reliable, and seems to have no weak points. One of our local drivers, John Porter, put 700 racing miles on his before opening it up. About all we ever do is change plugs. At the last Sacramento race it's true that I threw a rod, but this came as no surprise. The engine I was using had started out in life as an 1100 cc. It did a season of racing in Europe, including Nürburgring and Le Mans. Then it came over here and ran at Sebring. After that it got a set of 1500 cc barrels and raced a season in the east. Then we got it. We raised the compression, put the valve spring pressure up, and substantially over-revved the little engine in three races. It hadn't been apart for three years and we knew it was due to blow any day, but it was running so nice and loose that we didn't want to tamper with it. It had a perfect right to let go when it did."

The Spyder engine is one in which everything is adjustable and consequently it's quite easy to get something out of adjustment. But it's a simple engine to work on, provided that you take care to put it back together as original. One of the bugs of the early engines was the tendency of the intake valves to lose their clearance. This was due to hammering of the bronze valve seats, now replaced with stainless steel. There is an inevitable passion among race-car owners for more and more engine revs, and as soon as one owner starts over-revving, the others are likely to, too. The Spyder engine's revving ability can be easily raised by putting shims under the valve springs. The result is that today most of the front-rank Spyders are being wound 1000 to 1500 rpm over the factory limit. This imparts to the engine a certain degree of fragility that the factory is in no way responsible for.

The Spyder's brakes are normal Porsche assemblies with two leading shoes and two cylinders at the front, and one trailing, one leading shoe, and one cylinder at the rear. Their servo action is unusually strong, and learning to use these brakes with the right degree of delicacy is one of the tricks of learning to drive this car. The brakes are always there; there's never any doubt about them. However if you lay a heavy foot on the pedal and just hold it there you'll lock the brakes for sure, and this, Ginther feels is the most common reason that Spyders spin out. The brakes on his car had used up all the pedal travel at Palm Springs, a few days before. This was the first time, braking hard from 125 mph or so, that Ginther had observed any brake fade on a Spyder. Before we took the car out, its brakes were adjusted and when we tested them they felt like new. With three races already on them, Ginther felt sure that the linings, which have a tremendous area in relation to the car's weight, would be good for another six meets of Pomona or Paramount Ranch calibre.

Spyder is jacked at reinforced point riveted to underside of skin.

In the article on Porsche engines in last June's SCI we pointed out that for better weight distribution the 550 engine in the Spyder is mounted ahead of the transmission instead of behind it, as in all other Porsche models. The current Spyders use a normal Porsche-patent four-speed box with synchomesh on all gears. A nice feature of this transmission is that the ratio of each speed can be altered without affecting any of the others. This makes it possible to adjust each ratio in the box to suit the requirements of a given circuit—very handy when you're dealing with an engine with a narrow useful rev range. Even in standard form top gear is indirect and you can equip it with any pair of gears that have a total of 49 teeth. Miles keeps detailed records on the gearing requirements of each turn on each course he drives. When he's due to drive there again, he checks his notes. If Second was too low and Third too high, he can tailor the gearbox to suit the course. This is a more flexible arrangement than just changing the ring and pinion.

For such distance events as Mexico and Le Mans, Spyders have used a 4.43 ring and pinion and 0.88 top gears giving 125 at 6700 and a race "cruising" speed of 135 to 140 mph. This amounts to an overdrive, and the normal Fourth ratio of 0.96 can then be used for Third. The normal Third of 1.23 is used for Second, and normal First is retained as a starting gear. The new space-frame Spyder has a five-speed all-synchro box and 4.85 to one final drive. Ratios are:

I: 11/35 . . . 3.18
II: 15/32 . . . 2.14
III: 19/28 . . . 1.48
IV: 23/26 . . . 1.13
V: 25/24 . . . 0.96

For some reason it seems logical to pair the Spyder with the 300 SL. Both are dual-purpose cars, that can be raced or toured. Both are engineered with boldness and imagination, and both are leaders in their sphere. Certainly what the 300 SL is to its fast and elegant class the Spyder is to 1500 competition machines. Because they cost about as much as a basic 300 SL, few Spyders have been bought for touring use alone, and cost will probably keep it that way. Their real greatness lies in their competition potential, which they've proved by sitting on top of the 1500 cc racing world for three years now.

October's Palm Springs race, however, in which Bill Drake trounced the Spyders in a 1500 cc Cooper Climax, may have been an omen of things to come. Super-light Coopers and Lotuses are beginning to infiltrate the U.S., and good drivers are getting rides in them. It's true that they're sprint machines and perhaps would break up in a long, hard race that wouldn't faze a Spyder. But in the U.S. most races are short. Porsche Spyders here are going to have to work hard to hold their own. The Porsche factory will probably work out a few answers, and it's a cinch that the von Neumann stable will find a few of its own. Miles happens to have a new Cooper-Porsche, and maybe that's one of the answers.—*G.B.*

under the hood

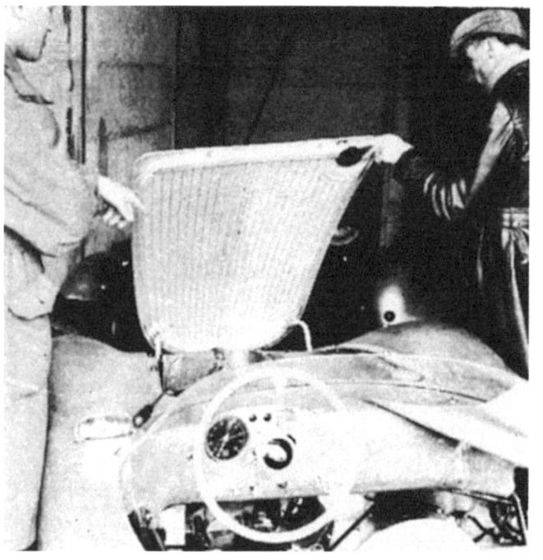

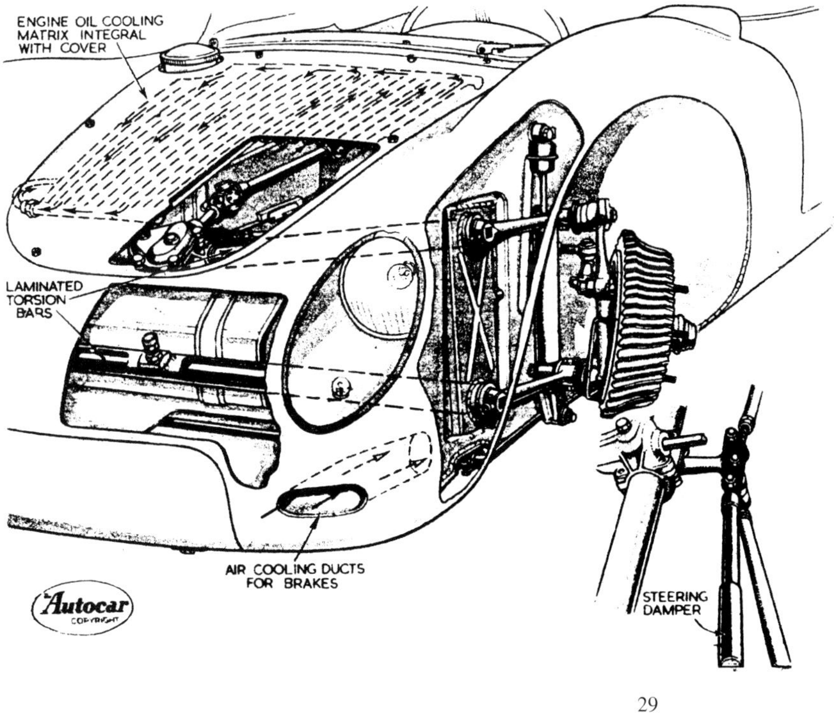

PORSCHE TYPE 718

Perhaps after reading Roger Huntington's article on page 40, this practical application of his go-fast principles will be of interest. First unveiled at the Nürburgring 1000 kilometer *Rennfest* in June, Porsche's newest Spyder RS2 is 26 pounds lighter and, shades of Detroit, nearly five inches lower than before, a 10% reduction in frontal area. The "trunk lid" is a surface type oil-cooler on hinges, which is about as ingenious as one can get, and a 5% improvement in penetration is achieved by eliminating the nose slot and cleaning up the neighboring sheet metal and plexiglas. Air resistance, therefore, is down to 0.855 (0.90 x 0.95) of its previous value.

German sources indicate that at its top speed of 149 mph (240 kph), the previous 1500RS, which commanded 135 PS (DIN, not SAE), had a ratio of air- to rolling-resistance of 6.5/1, splitting the 135 PS into 117 for air drag and 18 for rolling. Now, if the new body were installed around the old power unit, we would find the air drag reduced to 100 PS (0.855 x 117) and the total drag down to 118 PS at 149 mph. Since power required is proportional to the speed cubed, and assuming that suitable changes are made in the gear ratios, we can find the new top speed,

$$\left(\frac{V}{149}\right)^3 = \frac{135}{118} = 1.14 \text{ or } V = 149 \times 1.045 = 156 \text{ mph.}$$

This same seven mph increase would have cost 17 hard-to-find PS in the old body.

As a matter of fact, Engineer Hild and his talented team have managed to find about half that much. The cooling fan, pumping 1300 liters of air per second at 7300 rpm, consumes 8.8 PS, all of which will be available for go instead of blow when they perfect the exhaust-induced cooling-air flow installation. (This was pioneered by Fletcher Aviation in Pasadena on a Porsche-engined jeep in 1952-3.) For now, the unsightly bulge behind the cockpit (the world's largest headrest?) covers the usual Porsche shroud. Even more unsightly were the fins added for Le Mans' long "Muldoon" straight, but they didn't prevent Barth from spinning eight times in one go.

Now that the low-pivot point rear suspension is an accepted part of the racing Porsches, the factory is making changes to the front end, which Jesse Alexander details in his Le Mans report on p. 28. Stuttgart-Zuffenhausen is evidently profiting from the proximity of Stuttgart-Unterturkheim as the air is thick with fuel injection rumors, but nothing official has been said by the factory except "Go away."

1958 COMPETITION CARS

The PORSCHE Ty

Consistent Class Winne Contender in the Spor

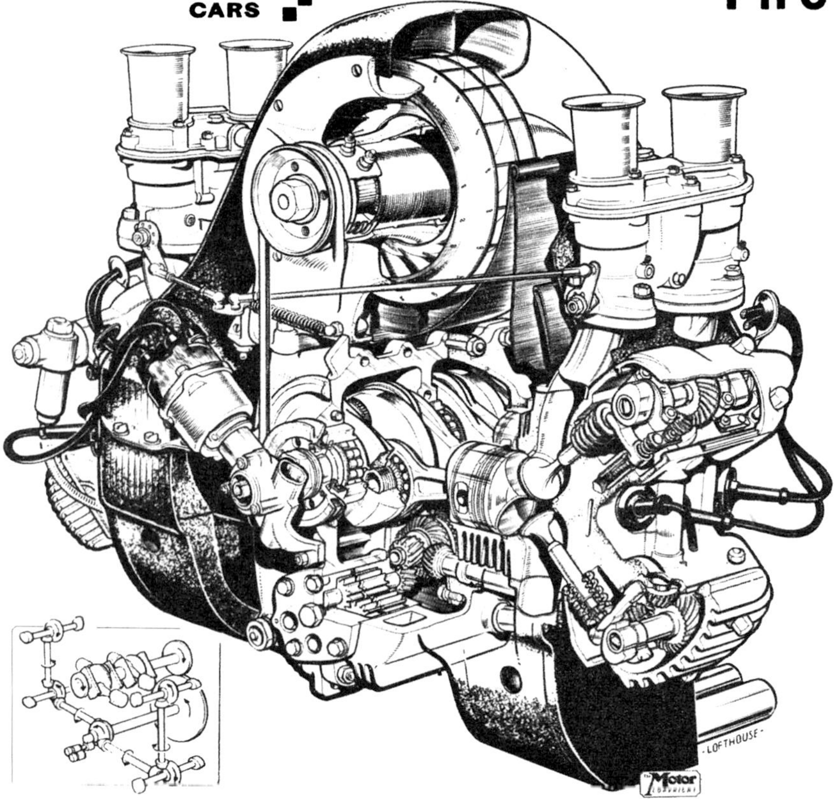

The 4 o.h.c. engine is bored out to 87.5 mm., raising the capacity to 1,587.5 c.c. Net output is 150 b.h.p. at 7,200 r.p.m., using new Weber carburetters and a compression ratio of 9.8:1. The camshaft drive and cylinder head layout are indicated in detail sketches.

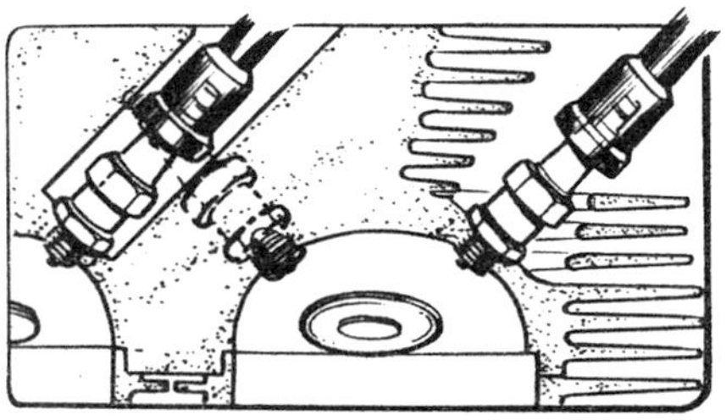

THIRD overall at Le Mans on 1.6 litres, fourth and fifth with 1,500 c.c.; thus the latest additions to the laurels of the racing Porsche. The squat two-seater, with or without tail fins, is not the most beautiful of sports-racing beasts now current, but on the basis of this year's Sports Car Championship alone it has a remarkable record. Championship marks after Le Mans leave it an easy second behind Ferrari, with a total of 20 points to Ferrari's 38, Aston Martin's 14 and Lotus' 3. It is when these are taken in conjunction with class and Index of Performance figures that one realizes how small a car has been responsible for them. Briefly, out of five races the Porsche has had 7 class wins, while against all comers it has taken one second place, three thirds, a fourth and two fifths, as well as a second place on Index. It is an interesting reflection that in sports-car racing, just as

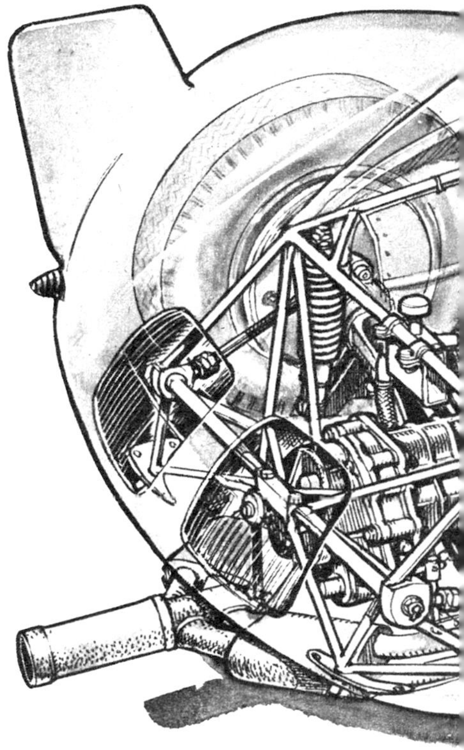

in the *Grandes Epreuves*, small cars are thus offering a direct challenge to the heavier machinery.

Both car and engine are products of steady development, rather than quite new designs, the racing department of the Porsche factory being very closely linked to production lines which every working day turn out some 25 cars of universally sporting nature. You can buy a Porsche in many stages of performance, almost up to that of the cars which are raced by the works, so competitions success has a direct bearing on sales.

If space allowed, it would be an interesting exercise to compare the latest design with the very first car produced by Ferry Porsche after the war, which has just been repurchased by the factory from its Swiss owner. Practically nothing now remains of the Volkswagen components and influence, save the use of trailing-link, torsion bar front suspension and an air-cooled opposed four-cylinder engine.

The heart of this unit is a light-alloy crankcase, split vertically upon the centre-

PORSCHE Spyder Type 718 Le Mans

Engine (1600)	
Cylinders	4
Bore	87.5 mm.
Stroke	66.0 mm.
Cubic capacity	1,587.5 c.c.
Valves	4 o.h.c.
Compression ratio ...	9.8: 1
Engine Performance	
Max. power ...	150 b.h.p. (DIN)
at	7,200 r.p.m.
Max. b.me.p. ...	183 lb./sq. in.
at	5,800 r.p.m.
Engine Details	
Carburetters ...	Weber IDM
Ignition timing control	Automatic
Plugs	Bosch W270 T17
Fuel pump ...	Bendix electric

Fuel capacity	28½ gall.
Oil filter ...	M & H 66/6. G72
Oil capacity	17½ pints
Cooling system	Air, ducted fan
Electrical system	Bosch 12-volt
Battery	24 amp. hr.
Chassis Details	
Brakes	ATE. 26.5, front
Brake drum diameter	11 in.
Suspension:	
Front	trailing arms and torsion bars
Rear	swing axle and coilsprings, with Watts linkage and radius arms
Shock absorbers	Fichtel and Sachs hydro-telescopic

Wheel type	Kronprinz 16 in. aluminium, pierced 5-stud disc
Tyre size:	
Front	5.00-16
Rear	5.25-16
Steering gear	ZF
Dimensions	
Wheelbase ...	6 ft. 10¾ in.
Track:	
Front	4 ft. 2¾ in.
Rear	4 ft. 1¼ in.
Overall length ...	12 ft. 5¾ in.
Overall width ...	4 ft. 10¼ in.
Overall height ...	2 ft. 10¾ in.
Scuttle height	2 ft. 3½ in.
Ground clearance ...	5 in.
Turning circle	36 ft.
Dry weight	10¼ cwt.

718 Le Mans

Strong Outright Championship

Latest and most successful, the Le Mans car driven by Hans Herrmann, who with Behra finished third in general category, ran without tail fins. All three Porsches broke the previous race records for their classes.

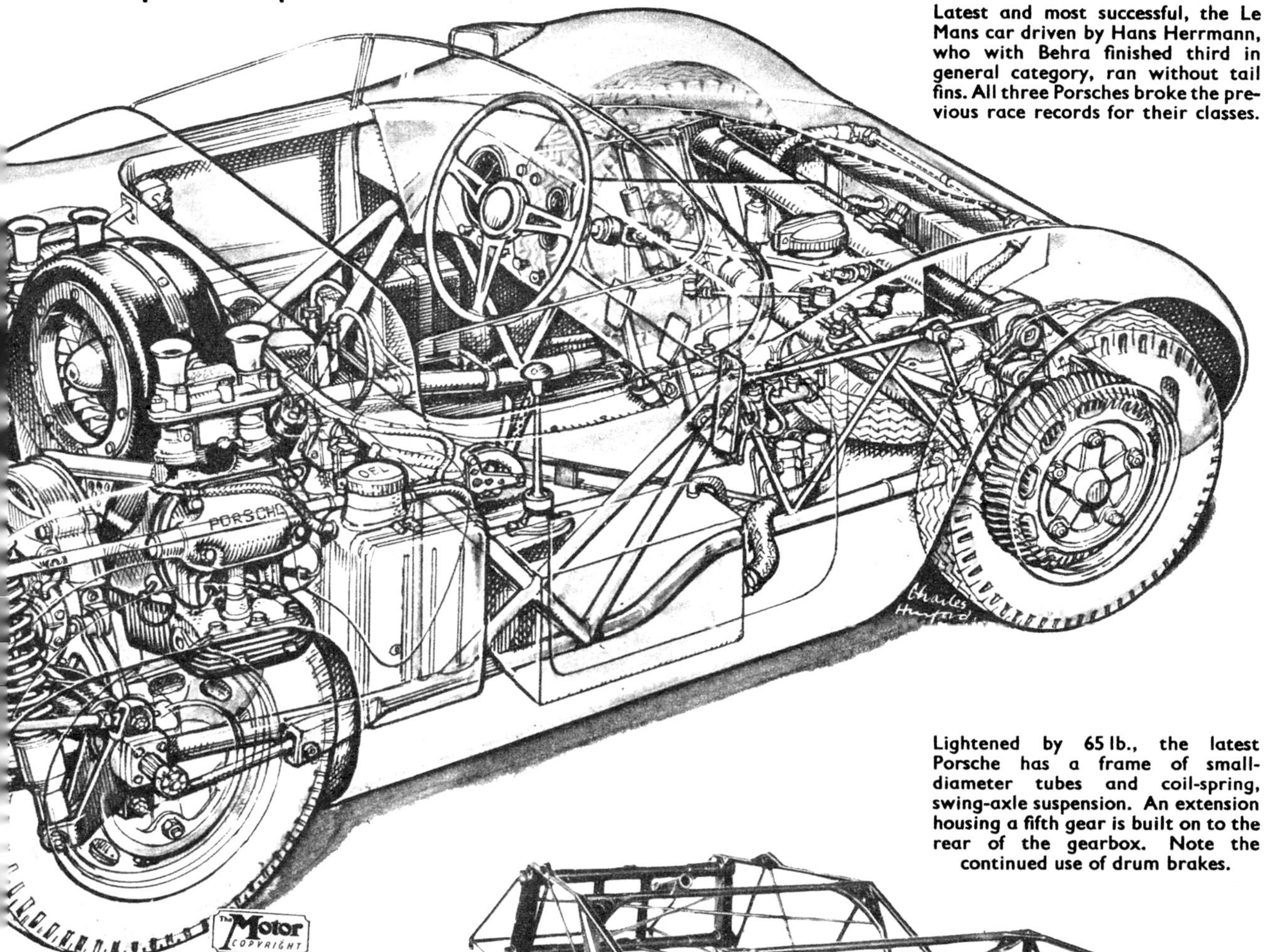

Lightened by 65 lb., the latest Porsche has a frame of small-diameter tubes and coil-spring, swing-axle suspension. An extension housing a fifth gear is built on to the rear of the gearbox. Note the continued use of drum brakes.

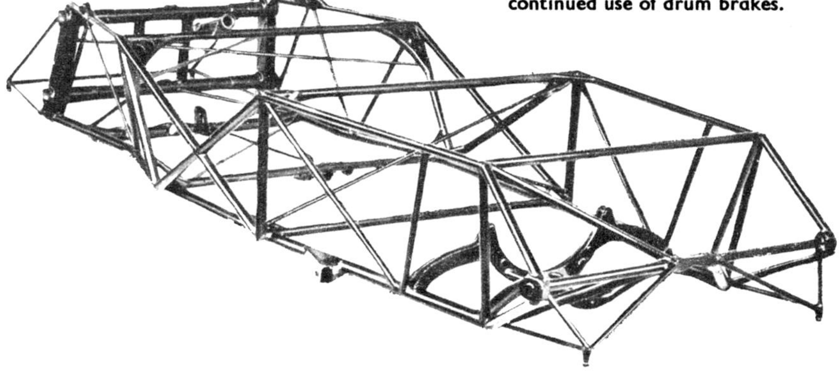

line of the crankshaft and carrying in its lower part a subsidiary shaft from which the camshafts and oil pump are driven. Rollers are employed for both main and big-end bearings on the Hirth built-up crankshaft, although there is a possibility of the racing engine following the line already taken by the touring models and going over to Vandervell plain bearings. Four long studs disposed around each cylinder pass right through the finned castings of the separate cylinder barrels, which are held down by a common head for each pair of cylinders. In common with production Porsches, the cylinder walls are chrome-plated, with a pattern of knurled "pockets" which retain oil upon the otherwise insufficiently porous surface. Successive increases in compression ratio to the present figure of 9.8 : 1 have restricted space in the hemispherical combustion chambers, so that the heads of the forged pistons are now very much recessed to clear the valves.

It was with the introduction of the now well-known four o.h.c. head that the Porsche engine made its greatest departure from the Volkswagen design. The valves are disposed at about 90 deg., operated through rocking fingers by a pair of camshafts on each cylinder head. Reference to the accompanying illustrations will explain more clearly than words how the drive is taken by spiral bevel gears and shafts to the two lower, or exhaust, camshafts, and thence upwards to the inlet mechanism. Very clean gas flow has been further improved by the adoption of new Weber IDM downdraught carburetters. A detail drawing shows the manner in which two sparking plugs per cylinder are fitted into the arrangement, the two sets being supplied independently by a pair of distributors on the crankshaft nose and coils mounted on the bulkhead. The net result of the apparently rather complicated design is a wide but otherwise compact unit running freely up to more than 8,000 r.p.m., for which 150 b.h.p. (DIN) is claimed as a conservative estimate—that is, a figure which can be held for at least five minutes. Flash readings of 160 b.h.p. have been obtained, but for circuits like Le Mans continuous performance is deemed more important.

Cooling of the engine is the function of

The PORSCHE Type 718

Roller bearings are used for mains and big-ends on the Hirth built-up crankshaft. The fine polish of the finished article can be seen.

a centrifugal blower, belt-driven from the crankshaft, directing air over the cowled cylinders and absorbing only about 8 b.h.p. at full speed, which must be appreciably less than the power used to propel a radiator through the air. Pressure and scavenge pumps in the crankcase take care of lubrication, in conjunction with an ingenious system of oil coolers. One cooler is of normal type, located on the cockpit floor and enclosed in an air duct, while the other consists of a series of flattened tubes on the underside of the front "bonnet" panel where there is naturally no obstruction whatever to the air flow. A thermostatic regulator bypasses both coolers until the oil is hot, while the reservoir for the $2\frac{1}{4}$-gallon supply is placed beside the engine.

Like all the Spyder models, the 718 has its engine in front of, and gearbox behind, the rear-wheel centres, mounting points on the clutch housing supporting the weight of the engine as a cantilever. The gearbox is a development of the VW component, carrying four helical gears (with Porsche synchromesh) very compactly in a magnesium alloy barrel housing while a fifth, or starting, gear lives outside the main casting at the rear of the box, in company with reverse. The gearbox is of the two-shaft all-indirect type with the advantage that all ratios can be varied independently. Three changes are in fact possible on each gear, while the bevels of the final drive will provide a ratio of either 7.1 : 1 or 7.3 : 1. The primary shaft is at the bottom, and the drive passes forward again from the secondary shaft to a ZF differential. The Porsche transmission is unusual amongst racing cars in relying on splash lubrication. Two breathers rise to a "header" tank above the gearbox.

The chassis is now a stage farther away still from the normal Spyder two-seater than was the car raced experimentally at Le Mans last year. A major change in suspension has permitted the tubular space frame to be lightened still more by elimination of the large-diameter cross member, which used to lie ahead of the engine, containing the torsion bars of the rear suspension. For the first time in Porsche history torsion bars have now been dropped in favour of coil springs for the rear wheels. Fabricated swing axles are pivoted on a cross member below the gearbox casing, while fore-and-aft location is now provided by the popular Watt's linkage, the radius arms having steel ball-joints at either end. The small amount of sliding freedom in the drive shafts, made necessary by the low-pivot design, was previously met by sliding joints in the shafts themselves, but Porsche have now adopted the alternative solution of a spherical pot-joint at the inner end of each shaft. This is a reversion in principle, though not in detail, to the VW. Trouble having been experienced with burning of the vital rubber covers which enclose the lubricant for the spherical joints, a spare cover is now fitted at the same time; if the cover shows signs of damage it can be cut away, and the spare slid straight on.

As in last year's Le Mans car the frame allows unusually deep spacing between the trailing arms of the front suspension, but the ball-joints which were an innovation at that race have disappeared again. The transverse torsion bars are retained, enclosed in the large tubular cross-member, but without the system of supplementing torsion bars which gave rise to the "K" in last year's designation RSK. Shock absorbers, as at the rear, are Fichtel and Sachs telescopic units. A small telescopic steering damper on each side is a noticeable reminder of one of the disadvantages of this geometry.

Tyre wear is evidently not considered a serious problem, for five-stud, bolt-on pressed aluminium wheels are employed even in long-distance races.

Development of the Spyder. On left, the production car; right, the RS with larger fuel tank and spare wheel moved to a front tray; in the foreground the car with stabilizing fins. The future of these has still to be decided.

In Britain, which supplies a greater proportion of the available makes of racing car than any other country in the world, it is easy to forget that elsewhere the disc brake is more an exception than a rule. The Porsche answer to this question is perfectly frank. They have experimented with discs, and others are currently experimenting on their behalf, but so far they prefer the drum brakes of their own design and manufacture, claiming that discs are 35% heavier and require five times as much effort from the driver. Whether this view is still held after the series of stops by Behra's car at Le Mans, in which first the linings and then a drum had to be renewed, has yet to be seen. Steel liners are fitted in the finned alloy drums, and the Energit lining material is

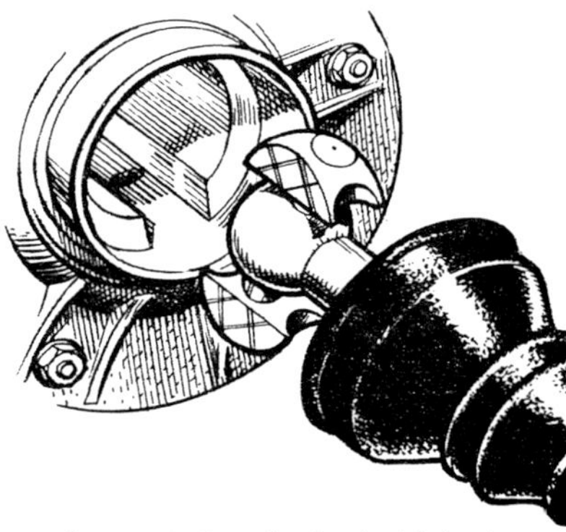

A new design of spherical joint in the final drive unit replaces separate universal and sliding joints formerly used with the low-pivot swing axle suspension.

an unusual mixture of steel wool and bunyl rubber. As a matter of historical curiosity the formula had its origins in a lining developed for the pre-war diesel express trains between Berlin and Munich.

With a large part of its weight placed inside the wheelbase, the 718 is naturally an extremely sensitive car to handle—more so than the Spyder, which may well be the reason for the experimental fins. In a high cross-wind at Nürburgring they were thought to be valuable, but only Paul Frère chose to have them for Le Mans. Weight distribution is slightly different from the production car, on account of the spare wheel which is placed right in the nose and the fuel tanks immediately above it. There is a supplementary tank beside the passenger seat, raising the total capacity to $28\frac{1}{2}$ gallons.

Spectators at Le Mans were able to observe that the handling of this extremely unorthodox racing car actually underwent a considerable improvement at a time when the road became treacherously wet. Until then, both 718 and RSK models had been cornering fast but untidily, but with the coming of rain the drivers were able to adopt a more conventional line. That three cars out of four completed the race (the fourth having crashed) testifies to the fact that the Porsche development staff have, like the drivers, covered more ground than most.

1960 PORSCHE RS60

Tailor-made for the Targa Florio

BY DEAN BATCHELOR
PHOTOS BY JOHN LAMM

"A SUPER CAR, beautifully well-balanced and simply tailor-made for the Targa Florio. . . . That was one morning I woke up and could say to myself, 'For today's race you have got the ideal car . . . ' "—Stirling Moss, talking about the 2.0-liter RS60 he shared with Graham Hill at the 1961 Targa Florio in his book *My Cars, My Career.*

Moss was not exaggerating, as his performance in that race quickly proved. The RS60 had started life as a 1500, but for this event the car had been fitted with 2.0-liter engine. Entered by Lloyd "Lucky" Casner's Camoradi (CAsner MOtor RAcing DIvision) but prepared and maintained by Porsche factory mechanics, the car had the measure of anything entered that day—at least when driven by Moss.

He set the tongues wagging after the first practice session when he lapped the 44.7-mile piccolo Madonie course in 43 minutes, 8 seconds in a 1.7-liter RS60, then brought his time down to 40:28 in a 2.0-liter team "mule." To put this in perspective, the best Ferrari time, by Wolfgang von Trips in a 246SP, was 43:42.

The race plan called for Moss to start, drive the first 4 laps, then turn the car over to Hill, who would drive 2. Then Moss, theoretically refreshed, would do the last 4. It seemed logical, →

as Stirling had more experience at that sort of racing (he won the Mille Miglia and Targa in the Mercedes-Benz 300SLR in 1955 and set the existing Targa lap record in an Aston Martin in 1958) while Hill's major experience had been in shorter Grand Prix races.

Cars started one at a time in the Targa Florio, at 30-sec intervals, with a minute and a half between classes. When Moss started the race had been underway for about an hour. Driving as if chased by devils (well, there was a horde of red Ferraris and Maseratis behind him), Moss moved quickly into 1st place on elapsed time, if not on the road. In his book, he described it this way: "I led for the first four laps by 1½ minutes from Jo Bonnier's sister Porsche and 2 minutes from the 3rd-place Ferrari, before handing over to Graham for his two laps. He handed the car back to me, minus 76 secs on the Ferrari, and I managed to change that deficit into plus 65 secs with one lap to go. That last lap was going together so well it looked like being a sub-40 minute record until—only 8 kms from the finish—the diff failed and put us out."

According to Henry Manney's report, his first for *Road & Track* and a classic of race reporting: "As the Camoradi car's fuel load lessened, Stirling accomplished the first of many lowerings of his own 1958 Aston Martin record. During the 4th lap Moss improved on his figure once again (40:58.2), even with stopping to hand over to Graham Hill.

"One would think that would secure a lead, but after a disastrously slow pit stop and EZ instruction to Graham Hill, the latter suddenly found Gurney breathing down his neck. Then, thanks to Hanstein's radio control, Graham was advised on lap six at the mountain station that Gendebien was leading him on time! And come in! Naturally he did so, rather feeling that he had been put into an ambiguous position, but in any case making up some of the lost time en route so that when Moss once again left the pit area, Gurney was just passing by.

"Moss driving or not, Gendebien's lead stretched out from 36.3 sec on the 6th to 1:1 on the 7th, but on the 8th a hurried

"With good luck you might set the timing in eight hours. It could take fifteen . . ."

visit for fuel to the mountain depot, plus the changeover back to Trips, cost the Ferrari the lead by 18.9 sec, although another Porsche record lap (40:41) by the balding Briton contributed its mite. By the end of the 9th . . . it looked as if Moss had pulled off another of his patented finishes."

Alas, it was not to be. After a superb drive by Moss in a superb car, the Porsche coasted to a stop about 8 kilometers from the finish line, its transmission dry because the securing bolts had stretched, allowing the halves of the case to spread and all the grease to escape. Nevertheless, the Porsche "Giant Killer" legend was growing.

The Moss Targa Florio RS60, serial 718-044 carrying race number 33, had run Le Mans in 1960, with Joakim Bonnier and Graham Hill driving. At that time it had a 1605-cc engine—a ploy by Porsche to move it into the 2.0-liter class, which allowed a larger fuel tank (100 liters, or 26.4 U.S. gal.). The reason for the only slightly larger engine was to retain reliability and allow more laps between stops.

This car also had the new high rear deck which, joined by the high side glass that faired into the new high windshield required by FIA regulations, made it into a sort of topless coupe. Porsche engineers had resented this new regulation that required a windshield at least 25 cm (about 10 in.) high, arguing that the increased wind drag hurt the smaller displacement cars more than it did the big-bore machines.

The 1960 race was not to be Porsche's year at Le Mans; engine failures put two factory RS60s out of the race; the third was 11th after suffering gearbox trouble.

The next appearance for 718-044 was at Sebring in March 1961. Now with a 1678-cc engine installed, it was driven by Herrmann and Edgar Barth, while a similar car was driven by Bonnier and Dan Gurney. The Bonnier/Gurney car had reached 9th place overall but retired just before the 7th hour with a badly slipping clutch. Team Manager Huschke von Hanstein called in the Herrmann/Barth RS60 and put the faster pair in, but it retired later with a broken camshaft, a problem that plagued more racing Porsches than the engineers would like to admit.

The 718 series Porsche Renn Sport had been introduced in 1957 as the 718 RSK. The spaceframe was basically the same as the previous 1500/500A but with a few tubes enlarged/deleted/added here and there. The main difference between the two models seemed to be the 5-in. lower and sleeker body of the RSK, a designation that came from revised front torsion-bar carriers in the shape of a supine K—a form of suspension that was fitted only to the first few cars of this type.

The wheelbase of the RSK remained at 82.7 in., as it had been since the first 550 Spyder, and the track was 50.8 in. in front and 49.3 in. at the rear—also as it had been from the first. These dimensions continued through 1959, but in that year the low-pivot swing-axle rear suspension was dropped in favor of a pair of wishbones on each side with an outboard Watt link for longitudinal hub positioning.

The RSK was an extremely fast and maneuverable car, but drivers complained about its twitchy handling. In a *Sports Car Graphic* interview, Ken Miles said, "If the RS was a taut car compared to the sloppy old 550, the K is tuned to piano wire tension, leaving very little margin for error. On a slick, low speed turn the car understeers horribly. . . . At over 100 mph the understeer resolves rapidly into an oversteer condition which must be corrected with very gentle movement of the steering wheel; any sudden or violent movement throws the car completely out of control."

Porsche engineers, cognizant of this, made some minor but significant changes with the RS60. The most important of these changes was a wheelbase increase of 4 in., to 86.7. This was to provide more predictable handling and more driver comfort (anyone but a small driver was cramped in the RSK). The rear frame could be unbolted, so that transmission work could be done without removing the engine and the wheels were now 15 in., with 5.50 and 5.90 tires front and rear.

Ignoring the wheelbase stretch, the RS60 body, built by Wendler in Reutlingen, looked much like an RSK. The major visual differences were a slightly blunter nose, driving lights set into the nose on either side of the air intake on the two Le Mans cars, and the FIA-required windshield, which Porsche mounted in a steel frame to make it look more like a streetable car. Racing experience dictated the mounting of an extra wiper—*inside* the windshield.

The FIA had also decreed that the cockpit should be at least 110 cm (a little over 43 in.) wide; the doors had to be at least 50 cm wide and 30 cm high (roughly 20 by 12 in.), and there had to be room on board for a suitcase measuring 65 by 40 by 20 cm (25.6 by 15.7 by 7.9 in.).

Meeting all these requirements wasn't easy for builders of small competition cars, but Porsche did it somehow. The suitcase space was found over the rear-mounted transmission, but fortunately an actual suitcase didn't have to be carried during the race—it would have interfered with engine cooling.

The standard engines for the first RS60s were the 547/3, giving 150 bhp DIN at 7800 from 1498 cc; 547/4, 160 from 1587 cc; and 547/5, 170 from 1678-cc. The 1966-cc engine in the Moss/Hill Targa entry produced very little more power than the 1.7, but the torque was up by 37 percent, a distinct advantage on that circuit.

The 4-cylinder, 4-camshaft engine in the RS60 was the work of Dr Ernst Fuhrmann of the Porsche engineering staff, who created the powerplant almost single-handedly. It ran for the first time on Maundy Thursday, 1953, three years to the day after the first production 356 rolled out of the Stuttgart shops. That first car's 1100-cc, Volkswagen-based pushrod engine produced 38 bhp, but the new 4-cam racing engine developed a whopping 112 at 6400 in its first test.

This marvel of complexity featured a built-up Hirth crank- →

136
CAMORADI

PHOTO BY BILL NEWROCK

PHOTO BY J.F. WELLINGTON

Top, at Le Mans in 1960 the RS60s had high-deck, semi-coupe bodywork. Before its restoration, 718-044 had "anteater" nose, above, used by Bob Holbert in 1961. Left, mechanics spent almost as much time building up Hirth crankshafts as drivers spent racing the cars.

shaft with roller-bearing mains and rods; four camshafts (two per side) driven by nine shafts, 14 bevel gears, and two spur gears; twin downdraft Solex carburetors; two inclined valves and two sparkplugs per cylinder.

The engine was not an assembly-line job; one man completely assembled each engine, a task that could take about 20–30 hours to build up the crank and another 90 to finish the engine. As Fuhrmann told Jerry Sloniger in an interview for his book, *Porsche, The 4-Cylinder, 4-Cam Sports & Racing Cars*, "With good luck you might set the timing in eight hours. It could take fifteen too, if tolerances weren't just right." Experience and newer techniques have reduced that time by half.

The front suspension was typical Porsche, with twin trailing arms on each side connected to transverse torsion bars. At the rear, the unequal-length double wishbones and concentric Koni coil-over shocks were continued from the RSK. The brakes were still drum type in those days, but enormous for the size and weight of the car: 11-in.-diameter finned aluminum drums with ribbed magnesium backing plates. A finished RS60 weighed only 1210 lb dry.

The first four RS60s were built as factory racers and differed from the customer cars in having separate right and left front torsion bars, instead of one long bar anchored in the center and springing both wheels. The factory setup can be identified by the two side-by-side adjustable torsion bar anchors, which allowed suspension tuning not previously available to the racing mechanics.

Twelve additional customer RS60s were built, seven coming to the United States, two to Canada, one to South Africa, and two racing in Europe. These cars had twin Weber IDMI carburetors and 9.8:1 compression ratios.

The Salon subject, RS60 number 718-044, was sold to Bernie Vehl, after Joakim Bonnier drove it at Mosport in mid-1961. It then passed to Hans Ziries and next to Clarence Catallo in Michigan.

When Warren Eads heard of the car being offered for sale in 1978, he was searching for any RS60 or 61. He had been through the Bondurant driving school in 1976 and after driving a 550, was looking for the ultimate spyder.

Eads flew to Detroit to look at the Porsche, which was in rather tatty shape, but when he saw the two front torsion-bar adjustments, he wanted that car. Eads had done enough research to know that this was an ex-factory racer. Further research revealed that Vehl had modified the body into the "anteater" configuration as seen when Bob Holbert drove the car at Nassau in 1961, finishing 2nd in national points to Roger Penske.

Eads chose Al Cadrobbi to do the mechanical restoration and Don Borth and Jack Hagemann to rework the body. Eads' RS60 is unique in one other respect; it has both the 1.7- and 2.0-liter engines that the car raced with in 1960–1961.

Laguna Seca is Eads' favorite track, one he used to visit during 1958–1962, hanging from a tree with binoculars or camera while watching Moss, Gurney, Jack Brabham, Phil Hill and others race on the then-new circuit.

Now he can race the same type of car he used to watch, on the same track. At the last Monterey Historic Automobile Races, Moss looked at the RS60 and talked with Eads about taking it to the Targa revival scheduled for this year. But having seen the Targa course, I later told him to forget it; that circuit is rough on cars and this RS60 is a living museum piece. Eads agreed; it should be kept intact.

Salon

1953 PORSCHE SPYDER 550-04

Porsche stakes its claim in racing

BY ROBERT T. DEVLIN & MICHAEL T. LYNCH
PHOTOS BY JOHN LAMM

ON MEMORIAL DAY, 1955, an obscure Sports Car Club of America race was held on a small airfield in Iowa City, Iowa. Two high school students from Kansas City, Michael Lynch and Fred Myers, drove up for the race. There was a certain piquancy when they found that the event's organization was so poor that they were able to enter without paying and stroll the pits with no passes. As they stood by a lone flagman, a few feet away from the intersection of two runways that was also the apex of one of the turns, the Salon car, the Porsche Spyder 550-04, flashed past. It was leading the under-1500-cc modified feature. The driver, Bob Davis, was followed by Ed Crawford in another Spyder. Davis won, and the two observers walked toward the finish line where the drivers were discussing the race.

A jubilant Davis addressed runner-up Crawford and Phil Stewart, who finished 3rd in an OSCA MT 4. "What are we going to do if people find out how much fun this is?" Davis asked. "Then everyone's going to want to do it."

Few in Iowa City could have known that 04's provenance included two Carrera Panamericanas and that it was Porsche's fourth purpose-built race car. Those first few race cars laid the foundation for Porsche's record as the preeminent marque in post-World War II sports-car racing.

Porsche's first works team raced 356 coupes in 1951 at Le Mans where they won the 1100-cc class. However, the German sports-car championship was won by

Walter Glöckler, a Frankfurt Volkswagen dealer, who built a group of mid-engine 1100- and 1500-cc Porsche-powered specials. OSCA, Borgward and the East German EMW were fielding thoroughbred racers in Porsche's displacement classes, and the factory realized that 356 coupes, no matter how heavily modified, were not going to get the results they wanted.

Porsche Design Group's project 550, directed by Wilhelm Hild, set out to remedy the situation. Although Porsche purists may claim that the inspiration for the 550 was the mid-engine Porsche Number One, the true precursors were Glöckler's specials. Porsche even had the first two cars bodied by Weidenhausen in Frankfurt, where the coachwork on Glöckler's cars had been done.

The 550 appropriately made its debut driven by Helm Glöckler, Walter's cousin. The event was the Eifelrennen at the Nürburgring in May, 1953. Porsche 550-01 was powered by an alcohol-burning Type 502 pushrod engine and won its class. The next month, two pushrod-powered cars with detachable coupe tops (550-01 and 550-02) went to Le Mans, bringing home a class victory. One also won its class at the Nürburgring in August, 1953. Shortly after, 01 and 02 were sold to a group led by a Czech-born, Guatemalan garage owner, Jaroslav Juhan.

Despite the successes of the first two cars, it was not Porsche's intent to make an upgraded Glöckler. Within the financial constraints of a small company, Porsche had a grand design. Concurrent with the development of the first two 550s, a young engineer, Dr. Ernst Fuhrmann, was working on Project 547, a racing engine to be built with Porsche's (and Volkswagen's) signature features of four opposed, air-cooled cylinders. Fuhrmann took inspiration from Project 114 in the Porsche design office's archives. It was a stillborn mid-engine sports car drawn by Karl Fröhlich in the Porsche office in 1938. The engine was a dual-overhead-camshaft V-10 with intake cams driven by bevel gears and shafts from the crankshaft, with other bevel gears and shafts driving the exhaust cams. Fröhlich's idea had also been used on the prewar Auto Union and postwar Cisitalia 360, both Porsche-designed. Fuhrmann's team designed and built a 1498-cc opposed-4 with classic dual overhead camshafts, dual ignition and hemispherical combustion chambers. Fröhlich's cam-drive system was used, and on the Type 547 engine, there were nine bevel gear-driven shafts. These hollow shafts carried oil to the camshafts and did not distort under torque or extreme heat. A Hirth built-up crankshaft was used, allowing for one-piece connecting rods. Assembly time was long, but once debugged, the engine proved to be bulletproof through years of development.

To house the new engine, two new cars were built, 03 and 04. These were much more finished and refined than 01 and 02. The bodies and frames were built by Karrosseriebau Weinsberg. Other than numerous detail refinements, the main difference in the chassis was the relocation of the rear torsion bars, allowing trailing arms rather than leading arms as on the first two cars.

The body was designed by Erwin Kommenda, who drew so many of Porsche's designs. His styling reach extends from the prewar Auto Unions, through the Cisitalia GP cars, the original 356 and 550, and into future Porsches, with the Boxster displaying 550 and RSK styling cues.

Later, 10 more pre-production Spyders were built. Cars 01 through 07 are distinguishable from one another because of body panels unique to each car, just like many Italian race cars of the period. Production Spyders, numbered from 0015 to 0090, were built by Wendler and differed in only minor details over their almost two-year production run.

Porsche had licensed an American company, Fletcher Aviation, to build Porsche engines for aviation applications in the United States. Fletcher sponsored 03 and 04 on a 1953 North American tour, which included a rare-for-the-time FIA-sanctioned race in Albany, Georgia, and the Carrera Panamericana in Mexico.

Spyders 03 and 04 had been extensively tested with the new 4-cam engine, but it was still not considered sufficiently reliable for the North American outing. Cam flutter was causing timing problems. This was overcome later by putting small flywheels on the rear of the offending 2-lobe intake cams to smooth out the reciprocal vibrations. In the meantime, 03 and 04 came to America with pushrod engines fitted. The outing was not successful, and both cars retired in Georgia and Mexico.

Car 02, driven by Guatemalan pri-

vateer José Herrarte, saved the day for Porsche in Mexico by winning the small-modified class that year, and Jaroslav Juhan used 01 to win his class in the January, 1954, World Sportscar Championship event in Argentina.

After the 1953 Mexican race, 03 was bought by John von Neumann, who was to become the Western VW and Porsche distributor. The car enjoyed a long career in U.S. racing in both 4-cam and pushrod form.

Our Salon car returned to the factory and reemerged a year later at the 1954 Carrera with the now reliable Type 547 4-cam engine. Hans Herrmann drove it to 3rd overall and 1st in class. This amazing performance in the epic five-day road race, combined with Herrmann's class victory and 6th overall in that year's Mille Miglia, established the Spyder's reputation as a giant-killer. The Mexican victory also spawned several series of Porsches designated "Carrera," a tradition that extends to this day.

There were several subtle styling differences between the bodies of 03 and 04, so their histories can be divined by examination of period photographs. The two main recognition points are the placement of the running lights (directly under the headlights on 03, inset toward the center of the car on 04) and a raised center section on the rear deck of 03. All pub-

The entrance to a Mexican motel served as a makeshift garage for both the 04 and 03 chassis at the 1953 Carrera Panamericana. The only woman driver in the 1953 event, Jacqueline Evans, poses with Karl Kling, who had won the 1952 road race in a Mercedes.

PHOTOS COURTESY JOE CAVAGLIERI COLLECTION

lished sources name 04 as the Karl Kling car in the 1953 Carrera and the Herrmann car in the 1954 event. Research for this article has established, to the authors' satisfaction, that 04 was driven by Herrmann in both the 1953 (race No. 160) and 1954 (race No. 55) Carreras.

After the 1954 Carrera, 04 went to the Nassau Speed Week races with Porsche sporting director and sometime race driver, Hüschke von Hanstein. As was the tradition with many European race-car manufacturers, redundant factory cars were sold in the North American market. Porsche 04 had served its purpose, and newer cars were being used by the factory.

At Nassau, a deal was made with Bob Davis, a Midwestern club race driver, that he would buy 04 on the condition that it would be prepared by the Porsche works mechanics before Davis entered the car at Sebring in March, 1955. It finished 10th overall and 3rd in class to two OSCAs. Davis kept the car throughout 1955, closing out the season at Nassau. It then was sold to Fred Vetter and spent the remainder of its racing career in the American heartland. One of its owners was Carl Haas, now the co-owner of the Newman-Haas IndyCar team.

When Tom Trabue bought the car from Norm Anderson in the late Eighties, Anderson had owned it since 1964. It was in original condition and had never been disassembled.

Trabue assigned the engine (90085) to Bill Doyle, who had set up a shop in Fresno, California, after training under 4-cam guru Jim Wellington of Rennsport Werke. Doyle brought the almost complete but tired engine back to period specification.

The chassis and body went to Joe Cavaglieri of Sherman Oaks, California. Most components were in good condition and only had to be freshened or occasionally replaced. The body was stripped, and a good deal of Bondo was found. The aluminum had hardened over time, but the gauge and methods of forming 90-degree bends and folds showed that the journeymen at Weinsberg probably had aircraft experience. The body appeared to have been formed over hardwood bucks. New form blocks were made to build the replacement panels. After a long, difficult job, German Siekens paint was applied. The team made reproduction decals so the car was as it ran in Mexico in 1954. Even a Carrera license plate, as given to all competitors, was fabricated to scale.

In 1990, with a car as beautifully restored as 04, Tom Trabue wanted the car seen where it could be appreciated. He took it to the Monterey Historic Automobile Races, where the world's most knowledgeable vintage-racing crowd marveled at the timbre of the car on the track and the perfect metal shaping and beautiful reproduction of the 1954 advertising graphics. Race organizer Steve Earle awarded 04 the Monterey Cup, given annually to the car "Judged to Have Excelled in Both Performance and Presentation." It was also invited to be displayed at the Pebble Beach concours the next day. Today, because of the foresight of both the factory and Tom Trabue, the car is in the Porsche factory museum.

Now in their 50s, Fred Myers and Michael Lynch were trackside that Saturday in Monterey when 550-04 once again flashed past them. For 10 laps that afternoon, their Iowa City outing, 35 years earlier, seemed like yesterday.

ROAD&TRACK
Radio

MEXICO
= 55 =
CARRERA PANAMERICANA
EFUNKEN
Radio

PORSCHE

ROAD RACER

Alan Hamilton's road-registered Porsche 908 is one of Australia's most exciting cars — a potent racing machine with a registration label affixed to its steeply-raked windscreen. David McKay drove the car through Melbourne's streets and around Sandown Park and came away impressed . . . hugely impressed.

PORSCHE

I CLIMBED out of my Avis Escort and there was the little wicked-looking white coupe lurking in the shadows of the building. It was a hot Sunday morning and Australasian Porsche distributor Alan Hamilton had kindly offered me the car to take to Sandown and back that day. The number plate told the story — AH 908 — by courtesy of the Victorian Motor Registry! To the factory back in Germany it was known as 908/0019, the last of the steel frame sports racers.

Chris Gribble of WHEELS was going to ride with me and even a quick peer into the cockpit told me he was going to be very uncomfortable. Chassis number 0019 was an but-and-out racer with no frills. To the average motorist the Escort's interior was lush in the extreme by comparison. There was no pretence at "finish" inside or out but the message was clear: this was built for a purpose — to run faster and longer than its rivals.

The roof line was lower than the Escort's waist and had I not had experience in inserting myself into tiny cockpits before I would have probably slipped a disc trying. The doors swing up and forward and open to a push on a button. The sill is wide for the driving position is near enough to central and the seat made of fibreglass (as indeed is the whole body) had a skin of cloth and looked uncompromising. In fact it proved remarkably comfortable.

Hamilton had the seat well back to accommodate his larger frame and he quickly found a pillow to jack me closer to the small black padded three-spoked wheel. I was adopting the attitude first made famous by that great Scot Clark and the Lotus 25, lying down to do the work. It was going to be very hot in such cramped quarters and under the big sloping screen. When raced, these cars carried ice up front and iced water was pumped through the drivers' suits at a flick of a switch next to the starter key. This combated fatigue in long races and was not bad thinking 12 years ago! Ventilation was minimal at low speed; the forced air was insufficient to keep the flaps open in the side windows and the only air came by way of the sliding panel in the rear window — along with the inspiring whine of the four chain-driven camshafts.

There was a neatly-fitted rubber mat around the pedals but good shoes, slacks and shirt are not advised for this sort of caper. I have become accustomed to fairly rough finish in race cars over the years but I have never seen anything so basic as the interior of the 908. The fibreglass mouldings around the tube sections were not pretty and I suppose I had been spoiled by aluminium-panelled cars of years ago. But everything worked and fitted reasonably well, though I was glad it was a fine day for I am sure she would have leaked like a sieve. The tits and bits are labelled in German and from left to right are: petrol pump, two ignition switches which look after the dual ignition distributor, a horn, a washer/wiper press-and-turn knob, light switch, the starter key and the aforementioned suit cooling switch and warning light carefully blacked over to allow no more than a pinprick of red light showing. Cowled ahead of the wheel are the vital clocks and warnings. Centre is the tacho (in this case a 906 counter redlined at 7800, whereas the 908 uses 8400), oil temperature which normally runs at 80 degrees is to the left and oil pressure to the right and indicates around 7 kg/cm at racing speeds. Clustered around the top of these clocks are two warning lights for the front and rear pads, one for fuel level, one for the alternator and one for oil pressure. This latter light comes on at anything under 4 kg/cm and so glows at idle and in traffic.

The pedals are well spaced and easy to operate, with a big left foot rest, and while the clutch is heavy by ordinary car standards it is no problem to dip once mobile. Naturally, the multi plates resent any fairy footing at traffic lights and prefer decisive movement and enough revs. The gear shift is beautifully placed to the right of the wheel and though first sometimes needs patience if stationary the other four gears, arranged in the proper H pattern, are as quick as you like. Reverse is locked out and has to be selected by a bowden cable, the lower of the two near the gear lever, the other being a choke and seldom used. This reverse cable did give a spot of trouble by coming undone and I was careful not to nose into any cul de sacs.

Hamilton gave me the cockpit drill, Chris folded himself alongside me somehow, we scraped over the gutter and off into Melbourne's Sunday morning traffic, snarled by some extensive road works. Driving a car which has to be worth the price of a good home unit is a responsibility at any time but when it belongs to someone else and everyone seems intent on bearing down on you, all you can do is relax and enjoy it as the actress said to the bishop.

Vision through the rear 180 degrees is not good. The interior mirror was mainly filled by Chris and I am never very happy about wing mirrors even if adjusted carefully. So one has to use the trafficators diligently (ahead of the wheel on a short stalk to the left) and rely on a sudden burst of acceleration if you intend changing lanes. Once we got to Dandenong Road and the traffic was moving relatively briskly, the 908 was happy enough in fourth slot. This kept the revs under 3000 and avoided the spitting and banging that heralded the coming on song over 3500.

The car was shod with Goodyear wets, 25 by 15 front and 27 by 15 rear on 9 and 12½ inch rims. For pottering, the pressures were 23/28 and I thought the ride over indifferent bitumen better than expected. The only fault was a tendency to pull right under brakes and obviously a caliper was sticking for I never had to use the parking brake on any gentle slope. This parking brake operates with the foot brake in the same manner as those foot-operated fly-off types so loved by American car makers. The small T-bar control hides ahead of the wheel to the right.

Naturally, the car caused a deal of interest

ROAD RACER

in the Sandown-bound traffic. Unlike some rare cars, there was no doubt as to its make; "Porsche" in the factory red lettering ran along both sides and across the tail while the famous shield was carried on the nose and aft of the doors.

If I were using such a car more than infrequently I would be tempted to fit a Ferrari roof periscope-type rear vision mirror and have 360-degree vision. The view ahead is superb: the two front wings channel the landscape in towards you and the road slips by underneath as though on a roller.

Certainly, there is a tremendous temptation to cut loose and run up through the gears, clearing both the plugs and the traffic congestion. A mild burst was sufficient to remind me that this is no super car needing to prove anything — it is a genuine racer and one which I have no doubt would win any long distance race here and probably any race over 500 km against the present crop of Group A cars. There is no great five-litre lurch forward when you open the throttle, rather more a sheer gathering of speed

like a bullet down a rifle barrel. Second and third gear ratios seemed fairly close, closer than expected, and fourth and fifth a little further apart.

I was able to trundle around the track at Sandown courtesy of Bill Patterson who has sole rights at race meetings and with the approval of the Clerk of the Course, John Redpath, and stewards. The tow wagons were busy clearing up after the sports sedans, so I was perhaps wisely constrained. I did manage one quick burst up the back straight, cutting off at 6000 in fifth (210 km/h) about mid way. (The factory recommends *above* 6000 when used seriously!) The Porsche's supple suspension with generous wheel movement was unperturbed by the notorious undulations of this part of the circuit.

It is only when you are able to enjoy the raw animal-like exhilaration of a true racing car that the time and money swallowed up in owning such a car are forgotten. It is a pity that real enthusiasts cannot line up one day and be given a quick ride in the passenger seat alongside Hamilton. Words can never adequately express the total satisfaction of feeling a fine factory racer becoming part of you as your right foot and two hands dictate the fine path between perfection and mediocrity.

Chris had asked me earlier would I like to own the 908 and I had said no for I know that I would be tempted to race it seriously and that would be foolhardy. But I did say later that I would rather own the 908 than any of the super cars of today. Regrettably, once you have driven out-and-out-racers the most exotic and glamorous road cars become a poor second best. So it might be better not to spoil the enthusiast's dream by letting him sample a taste of pure magic. Being a passenger is not the same as driving, whatever the car.

Chris and I drove back into Melbourne through the evening, content to allow the car to potter and concerned about returning it in good condition. I had noticed a small crack on the left front wing and feared someone might have leaned on the thin skin of fibreglass. Actually it had been inflicted at the Melbourne Motor Show when a small child was perched there for a photographer father.

The sinking sun was dead ahead at one point and there are no visors in race cars. One red-faced moment was near the St Kilda end of Dandenong Road. I stalled when taking off from the lights and there was no sign of life from the battery! I fancy we had inadvertently flicked the suit-cooling switch getting out of the car at the circuit and drained the juice. It meant a push start and Chris, with his arms full of briefcase and helmet, had to struggle out and apply one Gribble power to the 908 across the intersection until she caught and fired and then scramble back in again. Thankfully, the line of cars behind was patient and well mannered.

Back safely at the works, I thought about the car while Chris set off to ring Hamilton that we were home. I think the outstanding memory was one of efficiency, the car had made it an easy day. Some racers are cranky but this little German was so purposeful about everything. Being air cooled is a big plus when it comes to traffic driving — no boiling problems. And I was impressed by the hard, strong feel of the engine and its power/weight ratio. Hamilton told me that the best works cars had as much as 275 kW (370 bhp) and he reckoned this at around 260 kW (350 bhp). I have no reason to doubt him. He has a spyder version — 0018, which he intends putting together — but first there is the assembly of the 917/30, chassis 004. Hamilton now has all the parts and he is keen to have them bolted together.

I saw the last race of the Sunoco 917/30 with Donohue at the wheel at Riverside in October, 1973. He practised over four seconds faster than the second fastest car, won as he liked and retired on the spot. Hamilton says he should have 857 kW (1150 bhp) on tap but doubts whether it can be road registered. Now that should be worth waiting for, Mr Editor! *

WIDE EYES AND SMILES

By Chris Gribble

AT FIRST there is bewilderment. Then smiles, broad smiles. For some reason I'm still uncertain about, this mean, beautiful and purposeful racing Porsche brings joy to the strangest mix of people. Children smile, people who drive HR Holdens smile, even old people smile. Young girls are blase, though when you appreciate that they will rarely deem to even notice a car, a cool look is almost the same as a smile.

I know all this because I looked and learned from my foetal position in the passenger seat of Alan Hamilton's road-registered

NOTES ON 908 NUMBER 0019

THIS CAR came from the research and development wing of Porsche at Weissach. Built in 1968, it was the last steel-framed chassis. The 908/2 cars had alloy frames and there had been some experimental 910 and 907 models with alloy frames earlier. The Porsche book lists the 908 coupes at 660 kg but Hamilton's car weighed in at 630 at registration. Wheelbase is 2300 mm, length 4020, front track 1462, rear track 1454. (The 908/2 spyders were developed at the end of 1968 and were 100 kg lighter but suffered a higher drag penalty. The four hubs and the front stubs were of titanium. Because the 908/2 models were the first all-alloy frame models as a series, an air valve was fitted to the frame to pressurise the tubes and check for any cracks. Porsche continues this practice today.) Work was started in July, 1967, on the new three-litre eight-cylinder engine (908) and it was to drop into a 907 modified chassis, the 907 itself having been derived from the 910. The first 908 (001) was a short tail version and virtually a 907, including the 907 13-inch wheels. The first appearance was at the April, 1968, Le Mans trials. The first full 908 K (short tail) with 15-inch wheels raced at the Nurburgring that year. At Le Mans 1969 the long-tailed 908s, weighing 700 kg, were reaching 320 km/h. By the end of 1969 the 917s had matured sufficiently for the factory to race them and the 908s were sold

Porsche 908 while David McKay conducted us to Sandown Park one Sunday morning in April. I watched as people looked *down* on us from their family sedans as we slid below their normal level of perception. I watched as adult pedestrians momentarily forgot what was troubling them and smiled as we drove past. And I watched as young children shrieked their delight and boys on pushbikes did double-takes, smiled, craned to find just what this UFO was, then dropped their shoulders under the weight of their dreams. The seriousness behind the glassy eyes of teenaged boys on pushbikes still haunts me.

I can attest to all of the above. In the claustrophobia of the tiny cabin, with the urgency of the pops and crackles pouring into your ears through the small perspex slide window and, importantly, someone else driving, you can afford to look and listen. If you can understand, there is even a great deal of enjoyment in looking and listening. There is no question of smugness or exhibitionism — though they are both bountifully there if you want them to be. In my position as observer I put more value on the fact that I could look straight into the eyes of the people; by association, the object of their fascination.

Because David was driving, I could take more notice of drivers who pulled alongside us in traffic, tore ahead, then dropped back behind us to view the Porsche from all angles. I could also more easily take stock of our observations and the mental images which are conjured with the help of the wide eyes and smiles. I distinctly remember looking at a Solo service station sign and thinking: "Yes, that's the 908, so low".

The very fact that the car is so low can cause problems you don't appreciate in a sedan. The 908 was crotch-height to me, which translates to below the waist line of most cars. And drivers generally look straight ahead, not down. In the 908 you are obscured behind cars and vulnerable to people who are waiting impatiently to turn into your road from the left. In some close traffic situations you feel very vulnerable.

The stalling incident in Dandenong Road was illuminating in a way, in that none of our fellow traffic captives complained. We had stalled at the lights in the far right lane — a lane in which we were travelling because we were waiting to turn right a few streets on — and I had struggled out and given the car a push. Yet there wasn't a horn sounded or a word spoken in anger. I suspect the people were too awe-struck to remember their traffic tempers. Maybe I'm selling Melbourne drivers short. Though I suspect I was right the first time.

Even from the passenger seat, I could appreciate that the 908 was not an easy car to drive. Melbourne traffic is the antithesis of the car's intended function. At 6000 rpm, on a racing circuit and with a big heart the 908 would be a smooth and efficient car. On a city road, surrounded by looming cars and compelled to stay under 3500 rpm, it is an irritable and uncomfortable machine you tend to tolerate in traffic because it is unique, obviously potent and charismatic. I can speak highly of David McKay's composure and *sympatico* with the car.

Perhaps the most memorable aspect of reactions to the Porsche 908 was that it literally staggered the people who knew exactly what it was. David and I could have arrived at Sandown Park in a flying saucer and been no less noticeable. From the moment David poked our race passes through the tiny side window flap to the man at the gate the speechless admirers came from the woodwork. People literally stopped in their tracks, then gathered around as David brought the car to a halt underneath the grandstand. We clambered out to a reception committee of people who didn't know us. And every last one of them was smiling. *

off to private owners. The most successful of these was Reinhold Jöst, who continued to campaign 908s in various guises and only as recently as March 16, this year, started from pole in the Brands Hatch Six Hours and led in his 908/3 2.1 turbo (335 kW) from the Siegfried Brunn three-litre 908/3 and would have won comfortably but for a gear linkage breakage. Brunn incidentally has made contact with Hamilton and offered him 140,000 dms for 0019. Who says gold is the best buy! The factory developed the 908/3 for specialised races such as the Targa Florio and the 1000 kms at the Nurburgring, circuits where the bigger, more powerful 917s might not be the answer. These 908/3s used 13-inch wheels and weighed only 545 kg with oil. The complete body weighed a mere 12 kg. The 908 engine took four months from drawing to assembly and gave 310/320 first time up. For the first race at Monza the engine developed 335 bhp and this was to reach 370 in final tune. The whole engine weighed 178 kg. Bore and stroke were 85 and 66 mm, giving 2997 cm^3. The four camshafts were chain driven and the valves were sodium filled. Compression was 10.4:1 and torque was 294 Nm at 6600 rpm. The original transmission for the 908 weighed in at 25 kg and this was considered excessive! Ultimately, a new box (type 916) with five speeds (not six) was developed with pump-operated jet lubrication. Engine vibration was cured by using a flat crankshaft. *

The Hermann/Larrousse 908 leads a similar car out of the esses at Le Mans, 1969. 908s were reaching 320 km/h during the race that year.

Spin in a Spyder

Throttle response is good even when the engine's peak-power operating range is reached on the straights

by GRIFF BORGESON

with KEN MILES and RICHIE GINTHER

At Le Mans last year, a Porsche Spyder, driven by Umberto Maglioli and Edgar Barth made its appearance for the first time. The car, which was distinguished by small stabilizing fins at the rear, was in the lead in the 1,500 c.c. class before it tangled with a spinning Aston Martin. After this disappointment, further development was carried out in the car and a modified version was entered for Sebring in March of this year. In this article, *S.C.I.* correspondents give their impressions of the "ordinary" Porsche Spyder.

THE Porsche Spyder is a fantastic machine. It is fast and fiercely powerful, one of the most successful competition sports cars in history, and it has a personality no less complex than Dr. Jekyll's. At brisk highway speeds it is very easy to drive, docile, perfectly manageable, superbly responsive. It is even an easy car to drive "fairly" fast, the way Ken Miles and Richie Ginther have driven it in a long series of spectacular 1,500-c.c.-modified wins. But to drive it really fast, as fast as it's able to go—this is work for a driving genius—for a man, as Miles puts it, "with extraordinarily quick reactions and an exceedingly delicate bottom." Says Miles: "I'm just not that much of a driver."

At the same time there is nothing in the car's character that makes it inappropriate for use as a town hack. It has enough road equipment to satisfy those who do not like their sports cars too austere: hood, windscreen, trafficators, speedometer, wipers, and there's nothing to keep you from having radio and heater installed. Miles has a Spyder that he uses daily in city traffic, and he says, "Why not? Set up for town driving, with soft plugs and the right jets, it's a joy to drive—comfortable, roomy, and it goes like a bullet from a gun when you open the throttle."

I can vouch for part of his description—the last part. But *S.C.I.*'s test car, Ritchie Ginther's hurtling number 211, was hardly in town tune. The hard plugs were still in it from Palm Springs. The ¼-in. bored out intake ports and the whopping dual-throated Webers, mounted in place of the standard Solex carbs, reduced low-r.p.m. fuel-air velocity to a near minimum. In the interest of better weight distribution, Ginther has gutted the cockpit and moved the spare tyre and wheel in from their normal position aft of the engine. Sharing the passenger side with these impediments made the word "roomy" no longer accurate. With the big red numbers on the side, Ginther's Spyder didn't even *look* like a town machine. And there was, of course, the exhaust.

In stock form the Spyder is fitted with expansion chambers in the forked exhaust pipes and a perforated can in the single tailpipe, and it's a reasonably quiet machine. But with a set of large diameter, straight-through pipes the sound is shattering. A single 4,000-r.p.m. blip on the throttle and you expect to see cracks spreading in nearby concrete walls. The exhaust note starts as a resonant, thundering hammer and builds to that indescribable scream that can come only from a barrel chested 7,500 r.p.m.

It was in this conspicuous fashion that Ginther and I took off through Hollywood in heavy morning traffic, in an aura of fragrant fumes. To reach the coast in order to bear north, we had to pass through Beverley Hills, a town that takes pride in the strict enforcement of its anti-noise laws. It's a handsome tribute to car and driver that we ran the gauntlet without accident. The Spyder engine stroked along docilely at

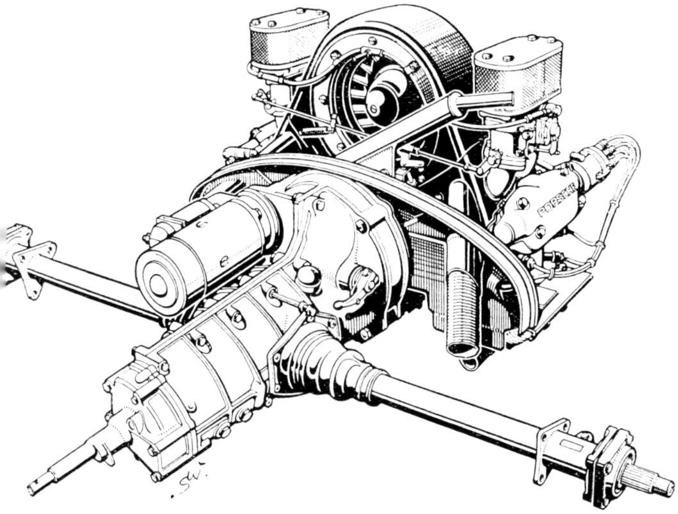

Spyder engine complete with transmission and final drive unit. Dual choke Weber carburetters are optional in place of Solexes

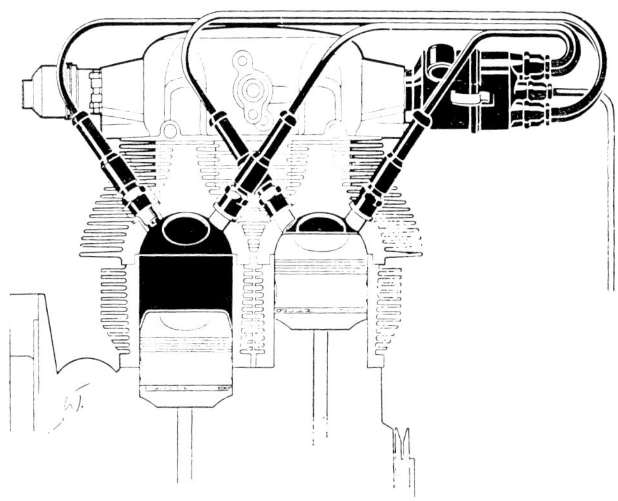

With two plugs per cylinder firing simultaneously the chances of misfires are eliminated and fuel from both ends of the chamber is ignited

1,500 r.p.m. in top gear! In spite of its steep power and torque curves it felt enormously strong at low revs, strong enough to make other 1,500 c.c. road machines seem feeble by comparison.

Once we reached the open country, though, we could give the beast its head. The first manoeuvre was to take it through the first three gears up to 7,500 r.p.m., to clean out the residues of an hour's crawl through traffic. The savage, lunging character of the car under full throttle is unforgettable. Up to about 5,000 r.p.m. it feels like one of the thrustiest machines you've ever driven, but then the cams hit their stride and the power *really* comes on. The most useful power range is concentrated in a narrow band between about 6,000 and 7,000 r.p.m.

"Geared for 125 m.p.h."

"It's hard," said Ginther, "for a guy getting out of a pushrod Porsche to learn to keep the revs up. In a normal Porsche you get up to 5,500 and the valves bounce. A pushrod man in a Spyder invariably feels that he's starting to wind the engine pretty tight long before he gets there. This car is geared for about 125 m.p.h. at 7,500, and I take it to 7,800 regularly.

"At Palm Springs, I tried to keep it above 5,500 coming out of all the corners. There's a 1,400-yard straight on the course, and if I came out on to it clean and fast at about 80, I reached a good 8,000 r.p.m. before I had to shut off. We know we can do 8,200 without valve gear trouble, and even at this figure piston speed is no problem."

Learning to drive by the rev counter and ignore the scream of the engine may be tough, but it's childishly simple compared with taking the car around a race course at something approaching its limits of adhesion.

"The Spyder," Miles says, "isn't like, say, a Maserati 1500—a car that anyone can get into and drive right on the ragged edge. With an ordinary car, when you're cornering fast there's a point where the car is perfectly stable, a point where the car is gone, and a substantial transition point, where you know it's starting to go and can do something about it. With the Spyder it happens *now*. A fairly good driver has no trouble taking it up to the point where it's nearly gone—but the brilliant driver, the genius, can add 5 m.p.h. to the cornering speed beyond that point. I'm referring here to a Kling or a Moss, not to myself."

You'd never, of course, discover limits on the road. At speed you can't even detect that the Spyder has a swing axle. It rides dead-flat and stays glued during terribly fast cornering. In a 30-mile ride over twisting mountain roads in Ginther's car, I noticed no sign of that usual "Porsche characteristic"—the slippery rear end. There's a lot less change of track in the Spyder than in the ordinary Porsche. Its very considerable negative camber brings the pivot point of the swing axle closer to the road. A line drawn through the point of tyre contact with the road is more nearly parallel with the road, so that the contact points move more nearly straight up and down.

You do notice the swing axle on a fast corner with an uneven surface, where you get great side forces combined with a lot of wheel movement. This, added to the fact that the front suspension flexes a lot under these circumstances, makes the car a hell of a handful on a fast, bumpy course. The unsprung weight in the rear is fairly high in relation to the weight of the car; it includes the big axle tubes, brake drums and radius arms. Among the effects of all the weight is a good deal of rear wheel patter when you're braking hard on rough surfaces. A lot of this has been eliminated in the Spyder, which has a separate tubular chassis. The frame is a much

Not so stark as some competition cars the Spyder has direction indicators on the steering column. Pedal arms are drilled and frame tubes are exposed to view

PORSCHE SPYDER

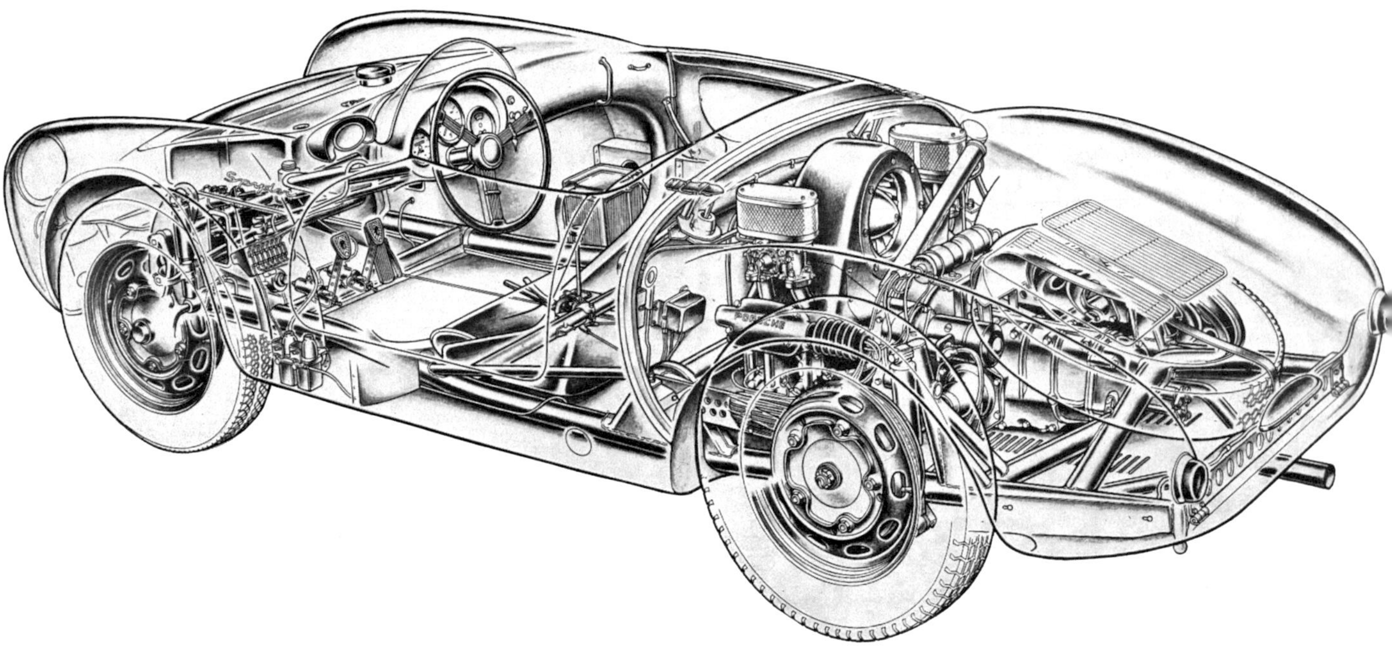

Copyright SPORTS CARS ILLUSTRATE

ENGINE

Type	Four-cylinder opposed air-cooled
Valve arrangement	V-inclined valves. Twin o.h.c.
Bore and stroke	85 mm. (3.35 in.) by 66 mm. (2.60 in.)
Capacity	1,498 c.c.
Compression ratio	12.0 : 1
Carburetters	2 Weber double choke
Power output	130 b.h.p. at 7,000 r.p.m.

CHASSIS

Wheelbase	82
Track (front)	51
(rear)	50
Suspension (front)	Transverse torsion bars (laminated)
(rear)	Transverse torsion bars (solid)
Shock absorbers	Telescopic double-acting
Steering	ZF-Gemmer worm and ball
Brakes	ATE hydraulic; two leading shoes front. Total area, 165 sq. in.

TYRE SIZES

Front	500 × 16
Rear	525 × 16

GENERAL

Length	141 in.
Width	61 in.
Height	40 in.
Weight distribution, front/rear	48/51

PERFORMANCE FACTORS

B.h.p. per litre	86.6
B.h.p. per sq. in. in piston area	3.69
Piston speed at max. b.h.p.	3,033 ft. per min.

Ritchie Ginther explains the workings of the big downdraft Webers. The rear dampers are enclosed under stressed sheet housings which supplement the original large tube supports

Spyder gearbox utilizes a Volkswagen case and offers a wide choice of ratios. Synchromesh is provided on all four gears

Piston crowns are deeply bevelled to provide clearance for valves. Symmetrical cylinder finning helps prevent uneven expansion of bore

Twin plugs are placed near valves and seats to minimize local hot-spotting while their wide spacing and angle ensures complete combustion

stiffer structure which helps the front suspension, and the rear suspension *always* has negative camber.

Of the Spyder's handling qualities Ginther has this to say: "The fastest way around a slow hair-pin turn is a matter of driver opinion. But on any fast turn the straightest, smoothest line is the fastest way. You can put a man in it who can dirt-track all over the course, but a straight-line man like Miles will drive right by him and not even look fast.

"If the Spyder's tail comes loose you can just leave it hanging out. You won't go as fast as when it's behind you, and I know that for sure. But you'll get into no trouble, and you can ignore it—which you'd better not do in the 1600 Speedster. Once the front *and* rear wheels start sliding, though, you're in a different situation. You have to anticipate how much room you have left, how much of it you're going to use, and manipulate throttle and steering accordingly."

The Spyder's steering is as light as they come. The brake pedal and clutch pedal pressures are light—very light. Throttle response is good, particularly when you approach the engine's peak-power operating range. There's nothing to changing gears; you shift with a true flick of the wrist. All the pedals are wide-spaced, and there's no doubt about which foot is on what. The ride is soft. There is no high-frequency noise audible in the cockpit. The screaming duet of engine and exhaust is carried away behind you. The result of these facts is that the Spyder taxes the driver very little. While a one-hour race can leave a Ferrari driver too weak in the legs to walk without limping and too weak in the arms to pick up his trophy, the Spyder leaves him with no more than a reasonable sweat.

For this reason, plus the fact that the Spyder is a heavy, solidly-built machine, it is difficult to contradict Miles when he calls the Spyder the finest long-distance race car in the world.

"A Spyder will go round and round a desperately bumpy course like Monza," he says, "and nothing will fall off. Meanwhile the lightweight racing cars in its class are likely to have broken in half. The Spyder is strong. For any long race—

for Mexico, Rheims, Le Mans, Nurburgring—nothing can touch it. On courses like these, the low driver-fatigue and even the high but narrow peak-power range pay off.

"The engine, if you don't over-rev it, is entirely reliable, and seems to have no weak points. One of our local drivers, John Porter, put 700 racing miles on his before opening it up. About all we ever do is change plugs. At the last Sacramento race it's true that I threw a rod, but this came as no surprise. The engine I was using had started out in life as an 1,100 c.c. It did a season of racing in Europe, including Nurburgring and Le Mans. Then it came over here and ran at Sebring.

"After that it got a set of 1,500 c.c. barrels and raced a season in the east. Then we got it. We raised the compression, put the valve spring pressure up, and substantially over-revved the little engine in three races. It hadn't been apart for three years and we knew it was due to blow any day, but it was running so nice and loose that we didn't want to tamper with it. It had a perfect right to let go when it did."

The Spyder engine is one in which everything is adjustable and consequently it's easy to get something out of adjustment. But it's a simple engine to work on, provided that you take care to put it back together as original. One of the bugs of the early engines was the tendency of the intake valves to lose their clearance. This was due to hammering of the bronze valve seats, now replaced with stainless steel. There is an inevitable passion among race-car owners for more and more engine revs, and as soon as one owner starts over-revving, the others are likely to, too. The Spyder engine's revving ability can be easily raised by putting shims under the valve springs. The result is that to-day most of the front-rank Spyders are being wound 1,000 to 1,500 r.p.m. over the factory limit. This imparts to the engine a certain degree of fragility that the factory is in no way responsible for.

Neatly mounted, the oil tank is fitted with filler, vent, temperature pickup and connections to engine and front-positioned radiator. A felt-lined steel strap holds it in position

The Spyder's brakes are normal Porsche assemblies with two leading shoes and two cylinders at the front, and one trailing, one leading shoe, and one cylinder at the rear. Their servo action is unusually strong and learning to use these brakes with the right degree of delicacy is one of the tricks of learning to drive this car. The brakes are always there; there's never any doubt about them. However if you lay a heavy foot on the pedal and just hold it there you'll lock the brakes for sure, and this, Ginther feels is the most common reason that Spyders spin out.

The brakes on his car had used up all the pedal travel at Palm Springs, a few days before. This was the first time, braking hard from 125 m.p.h. or so, that Ginther had observed any brake fade on a Spyder. Before we took the car out, its brakes were adjusted and when we tested them they felt like new. With three races already on them Ginther felt sure that the linings, which have a tremendous area in relation to the car's weight, would be good for another six events.

For better weight distribution the 550 engine in the Spyder is mounted ahead of the transmission instead of behind it, as in all other Porsche models. The current Spyders use a normal Porsche patent four-speed box with synchromesh on all gears. A nice feature of this transmission is that the ratio of each speed can be altered without affecting any of the others. This makes it possible to adjust each ratio in the box to suit the requirements of a given circuit—very handy when you're dealing with an engine with a narrow useful rev range.

Even in standard form, top gear is indirect and you can equip it with any pair of gears that have a total of 49 teeth. Miles keeps detailed records on the gearing requirements of each corner on each course he drives. When he's due to drive there again he checks his notes. If second was too low and third too high, he can tailor the gearbox to suit the course. This is a more flexible arrangement than just changing the ring and pinion.

For such distance events as Mexico and Le Mans, Spyders have used a 4.43 ring and pinion and 0.88 top gears giving 125 at 6,700 and a race "cruising" speed of 135 to 140 m.p.h. This amounts to an overdrive, and the normal fourth ratio of 0.96 can then be used for third. The normal third of 1.23 is used for second and normal first is retained as a starting gear. The tubular-frame Spyder has a five-speed all-synchro box and 4.85 to one final drive. Ratios are:

To lift the Spyder, a jack is placed beneath the tough, reinforced point riveted to the underside

I.	11/35 .. 3.18	IV.	23/26 .. 1.13
II.	15/32 .. 2.14	V.	25/24 .. 0.96
III.	19/28 .. 1.48		

For some reason it seems logical to pair the Spyder with the 300SL. Both are dual-purpose cars, that can be raced or toured. Both engineered with boldness and imagination and both are leaders in their sphere. Certainly what the 300SL is to fast and elegant class, the Spyder is to 1,500 competition machines. Because they cost about as much as a basic 300SL, few Spyders have been bought for touring use alone, and cost will probably keep it that way. Their real greatness lies in their competition potential, which they've proved by sitting on top of the 1,500 c.c. racing world for several years. ★

Although still not elegant, the latest Porsche is much sleeker than its predecessor

The Porsche Type 718

The latest $1\frac{1}{2}$-litre sports-racing car also does duty as a Formula Two machine

THE controversial decision of the Commission Sportive Internationale to limit engine size to 1500 c.c. for Formula One cars after January, 1961, gives new significance to the design of racing $1\frac{1}{2}$-litre cars at the present time. During 1958 only five manufacturers, using four different engines, competed in international Formula Two grand prix racing. These five were Lotus, Cooper, Ferrari, Porsche and Borgward. A great deal has been said and written concerning the Coventry Climax engine which powers the Lotus and the Cooper, while the Ferrari Formula Two V6 engine, the Dino 156, was actually the forerunner of the present $2\frac{1}{2}$-litre Formula One engine. Surprisingly little, however, has been written concerning the Porsche Type 718 which forms the basis for the Formula Two car, or of the Borgward which made only sporadic appearances in road races but pushed Porsche very hard in the International Hill Climb Championship. Elsewhere in this issue we give some details of the Borgward. Here we deal briefly with the Type 718 Porsche.

The staggering performance of the two-seater Type 718 Le Mans car at the famous 24-hour race last year came as a culminating achievement after many years of success in this race. The car which came in third over-all was a 1·6-litre engined car and those that came in fourth and fifth were 1·5-litre engined cars. Apart from the difference in engine sizes, the two types of car were identical.

There were no radical changes in the Porsche power unit for 1958. The Type 718, of course, employs the twin overhead camshaft version of the four-cylinder, horizontally-opposed engine, based upon the original Porsche engine which in turn was derived from the Volkswagen unit. In the 1600 c.c. version (the exact cubic capacity being 1587·5 c.c.) the power output of the unit is 150 b.h.p. at 7,200 r.p.m. This is achieved on a 9·8 to 1 compression ratio. Although of apparently complicated design, the engine is extremely compact and of very moderate weight owing to the extensive use of light-alloy. The crankcase is constructed of this material and is made in two parts, joining vertically down the centre of the crankshaft and the centre of a subsidiary drive-shaft for the operation of the overhead camshafts. The drive is taken to the camshafts via a complicated series of bevel gears which operate first the lower, or exhaust camshaft on either side and thence, via a short vertical shaft, the upper or inlet camshaft. The combustion chamber is of hemispherical shape and the pistons are very deeply dished to accommodate the movement of the valves with the 9·8 to 1 compression ratio. Two sparking plugs per cylinder are provided and the latest Weber IDM down-draught carburetters are employed. A belt drive from the crankshaft operates the large size centrifugal blower which pushes air over the individual cylinders.

In general layout, the 718 closely resembles the earlier Spyder models. The location of the engine and gearbox is, of course, at the rear, with drive to the rear wheels, and it is so

Bolt-on wheels and drum brakes have been retained in the interests of lightness. The fins have not been used with any regularity

Bulges beside the headrest cover the intakes of the four Weber carburettors

Changes in the rear suspension have allowed the suppression of a frame member at the rear

positioned that the engine falls in front of, and the gearbox behind, the line of the back axle. Improvement in the body shape has resulted from the moving of the fuel tank and spare wheel further forward, and curving the front of the car considerably. Stabilising fins have been used in the back also, although not with apparent great effect and certainly without consistency. Like the engine, the gearbox is of Volkswagen derivation and is compact in the extreme. Five forward speeds are provided, the lowest being for starting only and situated outside the main body of the box in company with reverse gear. The gearbox is of two-shaft type and all ratios are indirect giving latitude in alteration of ratios. The drive to the rear wheels is via a ZF differential and low-pivot swing axle. A major departure on the rear-suspension layout was the introduction of coil springs for this model in place of the previously used torsion bars. A Watts linkage is used to locate the rear wheels. Sliding movement for the drive shaft, necessitated by the low-pivot layout, is accommodated by a pot joint and rubber sleeve.

These considerable changes in rear-suspension layout have been accompanied by a certain revision of the chassis of the car. Still of tubular space frame type, it has been considerably lightened and at no cost to the strength of the structure. A major contribution to the lightening process was provided by the elimination of the member which provided mounting for the rear torsion bars. At the front, suspension is by transverse torsion bars and trailing arms, slightly simplified from the previous year's model and using telescopic shock absorbers as at the rear.

The driver's headrest conceals the top of the cooling fan in an ingenious manner

The position of the power unit and the revised rear suspension can be seen in this illustration

Two curious features of the Porsche always strike race-goers: one being the use of bolt-on wheels, even for long-distance races where wheel change is a virtual certainty, and the other is the continuing use, in common with other European car-makers, of drum brakes as opposed to discs. The reasons for these two unusual items of specification in 1958 are that they have found by experience that bolt-on wheels are satisfactory for their purpose and they are not satisfied that disc brakes provide sufficient braking improvement to justify their very considerable extra weight which is, of course, proportionately more on a very light car.

The differences between the Type 718 and the Formula Two car are slight indeed. Owing to the rear location of the engine it is a comparatively simple matter to remove the steering position to the centre of the car and suppress the passenger seat. Indeed, for added facility Porsche are now making a simple conversion kit with the Spyder which will allow it to run as a Formula Two car. Of course when, and if, the new Formula One comes into operation the regulation will exclude the running of a fully-enclosed body.

PORSCHE SPYDER TYPE 718

Engine:	Horizontally-opposed four-stroke of 1600 c.c. (figures in brackets for 1500 c.c.)
Cylinders	4
Bore	87·5 mm. (85)
Stroke	66·0 mm. (66)
Cubic capacity	1,587·5 c.c. (1498)
Valves	4 o.h.c.
Compression ratio	9·8 to 1
Maximum power	150 b.h.p. (nett) (142)
at	7,200 r.p.m.
Maximum b.m.e.p.	16·2 m.k.g. (15·3)
at	5,800 r.p.m.
Engine details:	
Carburetters	Weber IDM
Ignition timing control	automatic
Plugs	Bosch W 270 T 17
Fuel pump	Bendix, electric
Fuel capacity	28 galls.
Oil filter	M & H 66/6 G 72
Oil capacity	$2\frac{1}{4}$ galls.
Cooling system	ducted air by fan
Electrical system	Bosch 12 V
Battery capacity	24 amp. hr.
Transmission:	
Clutch	single plate, dry disc
Gearbox ratio	Top 26 : 23; 4 23 : 26; 3 19 : 28; 2 15 : 32; 1 11 : 34
Chassis details:	
Brakes	Teves, front Duplex, rear Simplex
Brake diameter	280 mm. front and rear
Suspension	front torsion bars, rear coil springs with Fichtel and Sachs telescopic shock absorbers
Wheels	Kronprinz 3·50 D–16 aluminium
Tyres	front 5·00 by 16, rear 5·25 by 16
Steering gear	ZF
Dimensions:	
Wheelbase	6 ft. $10\frac{3}{4}$ in.
Track—front	4 ft. $2\frac{3}{4}$ in.
rear	4 ft. $1\frac{1}{4}$ in.
Overall length	12 ft. $5\frac{3}{4}$ in.
Overall width	4 ft. $10\frac{1}{4}$ in.
Overall height	2 ft. $10\frac{3}{4}$ in.
Scuttle height	2 ft. $3\frac{1}{4}$ in.
Ground clearance	$4\frac{3}{4}$ in.
Turning circle	36 ft. 0 in.
Dry weight	10 cwt.

SCI **TECH REPORT**

PORSCHE'S SIZZLING TYPE 718

By Stephen Wilder

"NOW LOOK, how can a half-size sports car be running second in the World's Championship? Why, it's nothing but an overgrown VW Special." Such remarks might get any bench racing session off to a good start, so let's look into it.

Obvious as its lineage is, the latest Type 718 Spyder is too highly developed by far to be referred to so casually in anything but jest. As to the question, well, part of the answer is Good Drivers, but it still takes good cars to get the good drivers. Another, to be sure, is that points earned by a car through stubborn reliability are just as good as those won by dazzling performance.

Analyzing the Spyder inevitably involves one in its design history. A couple of worthwhile engineering truths are strongly pointed up, too. For a basically novel concept to triumph against more conventional opponents, details and components of equal novelty must be worked out — successfully. The other is that compromises are as unavoidable in auto design as elsewhere.

Even the first Porsche is part of the Spyder's history. Back in 1948, its hopped-up Volkswagen engine was located in front of the gearbox, which moved the center of gravity forward at the expense of accessibility and usuable space. Except for its tubular frame and light alloy body, the rest was all Volkswagen, suspension, brakes, wheels and so forth.

Thus, from the very beginning, Porsches have all been linked to the basic VW layout; rear engine and gearbox, suspension at the front by trailing arms and transverse torsion bars and at the rear by swing axles.

Let's analyze this a bit. When the engine lies between wheels that don't steer, its width is no longer so important. There goes *the* big objection to the flat or opposed engine layout. As well, a much lower center of gravity is obtained. Incidentally, the Germans call them "boxer" engines as the plan view of the pistons in motion recalls the fists of sparring partners busily swapping punches.

With the gearbox bolted directly to the engine, it is more compact to have the rear axles in between the two. The final drive gears and the differential may then be installed in the same gearbox casing for lightness. The gearbox proper *must* be all-indirect with the torque path turning through 180°. One advantage is that any gear ratio may be altered individually. Disadvantages are that it is not easy to fit any sort of quick-change device and, though less significant, power losses will be suffered in all speeds, even top. But as a bonus, it's actually hard to avoid independent rear suspension with this arrangement. Swing axles are one of the simplest means of achieving it and their recent development leaves little to be desired in stability and control.

At the front, complete freedom of space allows the use of transverse torsion bars with paired trailing arms, but once these are chosen, a wide frame is called for. This is all right, since sports cars are quite wide at the cockpit and anyway, the flat four engine at the back is pretty wide itself. With such a wide frame, there's not much point in building a GP car unless you're going to go fast enough to need a streamlined body. This, to get a bit ahead of ourselves, is just where they were at Rheims this year. (First in the Formula 2 race.)

Now, given this very unusual layout with its myriad of novelties; once begun, it would have been impossible to turn the car towards a more normal configuration even if they'd wanted to. Not to be discounted with this firm or any other is that though the Racing Department races for the sheer love of it, Sales encourages them (when they're successful) for the advertising value to be had. The Engineering Department pitches in partly for the technical exercise but mainly for the useful lessons to be learned. Useful on the production line, that is.

One might venture to say that with all this VW background, "The breed improves racing", but the more normal statement is true too, as Porsche's road cars are continually showing up with some new device which cut its teeth on the Spyder.

Especially praiseworthy is that so much has been accomplished by Porsche despite their starting off on a completely different tack. It's not that they've found a new miraculous way to build cars better than everyone else's, for the Spyder does have its limitations. For one thing, the biggest they've ever run the engine is 1587½ cc and this is a long way from the FIA's limit of 3000 for Appendix C Sports Cars. And it takes a heap of rectangular Deutschmarks to beat that many centimeters. If they should feel it necessary to build such a big engine (big? – the size of a Studebaker six?), the large increase in weight would make them pause before they'd pop it into the rear of a stretched out Spyder. When you change one item so drastically in a well rounded package, you will have to change a lot of others if the package is to stay round.

The resemblance of the very first Porsche to the Spyder layout has already been mentioned. The next "pseudo-Spyders" were the Glöckler-Porsches. One of them, developing 85 bhp on alcohol, set three 1500 cc International records in 1951 at Monthlery before coming to the States in the hands of Max Hoffman. Though carrying four-pin knock-off disc wheels of BMW 328 pattern, it too had a tubular frame and an aluminum body and the engine was in front of the gearbox.

Opposite page, two Type 718s, with and without fins; behind them, an earlier RS Spyder similar to those seen in USA racing, Above, the Behra-Hermann 718, finless, speeds past less fortunate cars under the Dunlop Bridge at this year's Le Mans.

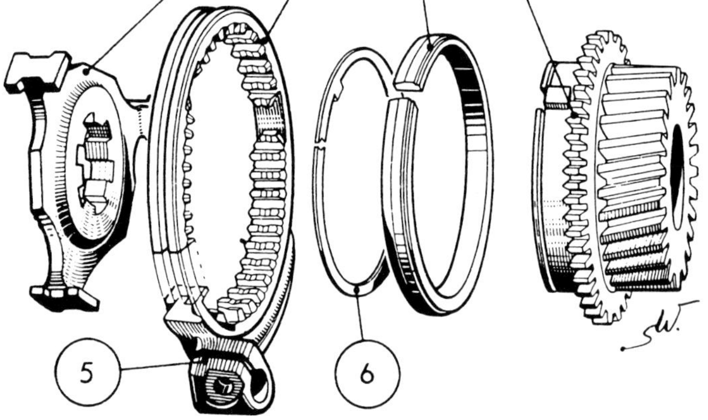

Exploded view shows Porsche synchromesh. Shift guide 1 is splined to pinion shaft. Operating sleeve 2, which turns with guide, is controlled by selector fork 5. Outer surface of slotted servo ring 3, tapered at left to match sleeve, catches at right beneath undercut teeth of dog clutch on driven gear 4. Safety ring 6 prevents ring from slipping off, square block on 4 keeps it from spinning. Text explains operation.

PORSCHE Type 718 Spyder

SPECIFICATIONS

POWER UNIT:

Type	Air-cooled, flat four
Valve Operation	Shaft-driven, four overhead cams
Bore & Stroke	3.45 x 2.60 in (87½ x 66 mm)
Stroke/Bore Ratio	.75/1
Displacement	97 cu in (1587½cc)
Compression Ratio	9.8/1
Carburetion by	Two Weber IDM twin-choke down-drafts
Max. Power	150 bhp (DIN) @ 7200 rpm
Max. Torque	117 lbs-ft @ 5800 rpm

DRIVE TRAIN:

Clutch	Single plate, dry disc
Transmission ratios available	
I	3.09
II	2.13, 1.94, 1.76, 1.61
III	1.47, 1.35
IV	1.23, 1.13, 1.04
V	0.96, 0.885, 0.815
Final drive ratio	5.17, 4.43
Axle torque taken by	gearbox case

CHASSIS:

Space frame of small diameter tubes	
Wheelbase	83 in
Tread, front and rear	51, 49 in
Front suspension	Widely spaced trailing arms, transverse torsion bars
Rear suspension	Low pivot swing axle, fore and aft Watts linkage, coil springs
Shock absorbers	Fichtel & Sachs telescopic
Steering type	ZF worm and roller, steering damper
Turning diameter	36 ft
Brakes	2LS front, 1LS rear
Brake drum diameter	11 in
Tire size	5.25x16 (5.00x16 optional at front)
Rim size	3.50x16

GENERAL:

Length	150 in
Width	58 in
Height	35 in
Weight, dry	1140 lbs
Fuel capacity	29 U.S. gallons

Porsche's first real race car, the modified 718's single seat, narrow headrest reduce frontal area for even better top speed.

Encouraged by Glöckler's success, but aware of his limited resources yet increasing competition, the factory built their first 550 Spyder in the spring of 1953. Work also commenced on a new flat-four air-cooled engine, to have two overhead camshafts per bank, but this wasn't ready until late summer. As it was, the Spyders were outstandingly successful with just the pushrod engines, winning several firsts and seconds that summer . . . Eifelrennen (Nürburgring), Le Mans, Avus, Nürburgring again and finally, the Freiburg-Schauinsland hill climb. The four cam engine first saw competition at this hill climb and, all too typically, finished third to the pushrod Spyder's first.

At first on the 550, as before, when the engine and gearbox were swapped around, the rear torsion bars were moved behind the axle line, too. But with longer flexible trailing arms (23.7 inches instead of 16.9), the torsion arms returned in midsummer to the more sensible location which creates toe-in on bounce and rebound rather than toe-out. Shock absorber geometry at the rear, none too satisfactory even on the VW, underwent several alterations, too. Eventually such a modication showed up as well on the A version of the Type 356.

In entering prototypes at LeMans, manufacturers once had to claim that these were the precursors of actual production models. Honoring this pledge, the Porsche works put the 550 Spyder on the market for a price (ex Stuttgart) of under $7000. At least 100 were sold, much to the discomfiture of the SCCA, but in truth, only a few were equipped or seriously intended for actual road use. Shortly after the supply was exhausted (no kidding, the factory was losing money on them), the works team appeared with a newer Spyder.

The RS (for Rennsport) differs in many details from the 550, but it still takes a careful look to tell them apart out on the track. A super-low starting gear was added on the tail of the gearbox, but the biggest step forward was in the weight saving and stability departments. A space frame replaced the simple ladder-like one and to lighten the body, the flip-top tail section was abandoned. Instead, two large access panels, very aircraft-like, permit changing plugs or carb jets but little else.

To improve stability through turns, the roll center was dropped by using low pivot swing axles. The drive shafts were splined to give sliding freedom. The spare tire was moved from above the gearbox — another detail from Porsche #1 — to the front end. At first strictly not-for-sale, these too trickled over to the States as soon as the works had something better underway: the Type 718.

But having come so far, making something better was no easy job and a lot of experimenting has been going on. Testing of the Type 718 went on during the entire '57 racing season and most of '58's. Early versions were known as the RS2 and then the RSK, the K indicating the angle of the upper front torsion bars. These allowed wider spacing between the trailing arms and at first ball joints were used. These have since disappeared. Known informally as the "Mickey Mouse" car, it was anything but stable at high speeds, von Frankenburg's crash at Avus and Behra's eight full spins on Mulsanne Straight (despite tail fins) at Le Mans being rather spectacular proof of it.

But this year's sweep of third, fourth and fifth overall at Le Mans indicate that earlier difficulties have been solved, so it is fitting to give a technical rundown on the current Type 718.

At Sebring this year, it was announced that coil springs were being used at the rear instead of torsion bars. A sharp break with Tradition (and the Sales Dept's point of view), it eliminated the heavy cross tube containing the torsion bars. The outer end of the low pivot swing axle is located by a fore and aft Watts linkage (see Technotes for August and this month). The sliding spline on the axle shaft itself is gone in favor of a spherical pot-joint at the inner end, a variation on the sliding U-joint used on the VW.

The gearbox uses the same barrel case featured on the production cars since late '56, the additional fifth gear being carried in a separate box at the end. Within the gearbox proper, the four ratios all use the famous Porsche "Servo-ring" synchromesh invented by Leopold Schmid. Though it can be "beat" in the automotive sense, in everyday language it can't — provided the driver makes an effort to get engine revs *somewhere* in the right neighborhood. Key to the ring's function is a gap which lets it be preloaded. Compressed slightly during assembly, it catches under a lip on the synchronizing dog clutch. Before the shifting fork can move the operating sleeve completely into engagement, the tapered inner surface of the sleeve must ride up on the similarly tapered outer surface of the servo-ring. As the pre-load is taken up, the frictional forces are sufficiently high to bring the servo-ring and the operating sleeve to the same speed.

Brakes are bonded aluminum alloy with steel liners and very fat transverse fins. Though extremely wide and capable looking, these seem to be a weak point, as lengthy pit stops at both Sebring and Le Mans have seen front drums being pulled. Linings by Energit are a compound of steel wool and butyl rubber!

One of the reasons stopping is such a problem is the splendid aerodynamics of the 718. The 10% reduction in frontal area from the RS to the Type 718 and the 5% improvement in penetration have been mentioned before in these pages. In addition, Porsche learned a lesson from EMW/AWE and now they, too, prepare the bodies as if for Bonneville runs.

The frame is some 68 pounds lighter than before. Made up entirely now of small diameter tubes, the greatest saving comes in the complete abandonment of the rear torsion bars and their encircling tube. The upper longerons are shoulder height, giving ample depth. At the cockpit sides where doors are required, X-bracing fills the necessary gap. At the rear, carefully shaped cross members cradle the engine and gearbox, while at the front corners, tubes are omitted in favor of stamped uprights which carry the front torsion bar mounts. It's an interesting reminder that a frame's only purpose is to tie all the *working* components together in a rigid fashion. And the simpler and lighter it can be, the better.

The front suspension is again by twin transverse torsion bars and paired trailing arms. The vertical separation is about 50% greater than on a VW, which should reduce braking loads on the trailing arms appreciably. The upper torsion bar tube is noticeably shorter than the lower one, but it's no longer kinked in the middle. The hydraulic shock absorbers by Fichtel and Sachs are fastened to the bottom trailing arm.

A drawback to the twin trailing arm layout is the impossibility of arranging steering linkage that does not cause toe-out on bounce and rebound (except with the heavy, complex Dubonnet system). This toe-out is no more welcome at the front than it was at the rear on the first 550, but it's harder to avoid. The divided track rods are unequal, which adds to the confusion. To prevent wheel shimmy, a small steering damper (miniature shock absorbers) is connected to the Pitman arm.

Placed underneath the lower torsion bar, the spare tire projects well forward, the body's rounded nose just curving out to fit. Above, the surface-type oil cooler has been built into what for convenience's sake is called the trunk lid. Actually, the space below it is completely filled by the gas tank. A supplementary oil cooler lies in an air duct along the cockpit floor; both are controlled by a thermostat regulator. An oil tank of about 2½ gallons is mounted in the engine compartment just behind the "passenger's" seat.

Dr. Fuhrmann's engine design features four camshafts which operate through light fingers to reduce side thrust on the large valves. The latter lie at about 90° to one another; so do the spark plugs, though they are offset from each other as well. The cam covers, the paired cylinder heads, and the individual pistons and cylinders are all aluminum. So is the crankcase. It splits on its vertical centerline to accept the built-up Hirth roller bearing crankshaft and a dummy camshaft. The latter drives the oil pump plus four camshafts and four intermediate shafts. All nine shafts connected by bevel gears. No doubt the adjustment of valve timing is not a casual job. Like the Porsche Super, the cylinder walls are chrome-plated and knurled. The artificial irregularities trap oil on the walls, while the chromium provides hardness the aluminum lacks.

When first announced, the Spyder engine released 110 bhp at 7000 rpm. Now the same size engine is poking out 140. With an extra 100 cc's, another ten horsepower is realized. These figures are all DIN and can be maintained for up to five minutes at a time. Over-revving is not easily catastrophic, 8500 having been seen on the tell-tale of cars which subsequently raced and lasted.

Dual ignition in the European sense is used — two plugs per cylinder and two separate distributors. The latter were originally mounted on the ends of the intake (top) cams, but now a bracket next to the end of the crank shaft holds them both in a V position. Carburetion is through two Weber IDM twin-choke down-drafts. No more has been heard of fuel injection since last year.

Engine cooling is of course by air. A belt-driven centrifugal fan of considerable aerodynamic quality does the job. Porsche engineers say it take only eight or nine horsepower to run it. Unsatisfied with even this efficient figure, they have been working since 1953 with a method of cooling which should reduce it to zero! This is not a joke, nor is it a variation on perpetual motion. The exhaust gas of an engine contains plenty of energy, you have only to hold your hand behind the exhaust pipe at full throttle to discover that. Why not use some of this wasted power instead of taking the hard-earned crankshaft power to blow the cooling air through? Compounding the engine, using an exhaust-driven turbine to run a compressor is a bit too complicated on so small an engine.

But it's on the right track. Patented over fifty years ago, Jet Cooling or perhaps more accurately, the Stream Pump provides the answer. Squirt a garden hose into a shallow puddle and what happens? The water already in the puddle is carried along by the fast moving stream. When this phenomonen is arranged inside a suitable boundaries, the efficiency is startling. Under certain circumstances, one pound of exhaust air can be made to carry along as many as fifteen pounds of fresh cooling air, an ample amount for even the hottest engine.

The drawback? Noise. Think how big a muffler would be required to handle sixteen Spyders at once! Though their engineers have been working on this phase of the problem too, there is a long way yet to go. So far they're said to be too embarassed to run it, even though it's a *racing* sports car.

There was a central seater Formula Two Spyder at Rheims this year. Are they ready to leap into Grand Prix racing? They already have the car, the drivers, and even the noise.

Stephen Wilder

PORSCHE Carrera GT

PORSCHES are indeed a different breed of sports car with their rear engine and air cooling. And the Carrera GT with its four overhead camshafts and dual ignition is a rather special sort of Porsche. Fitted with the same 128 hp engine once used in the factory's racing Spyders, we expected it to be the least road-usable of the many dual-purpose cars we'd been intending to sample for this issue. At first we'd planned to drive it to Lime Rock from its home at Frank's Auto Sales in Montgomery, N.Y., just to see how tractable it would be. Scheduling problems prohibited this but we were pleasantly surprised to see Frank Wagenhofer had caught the spirit of our adventure and brought it under its own steam himself. Usually it rides in serene splendor on a two-wheel trailer behind a shiny Mercedes-Benz 220S. A veritable Scuderia Stuttgart.

Frank confessed that most of the trip was done in second cog, with villages and towns along the route being traversed at 4000 rpm in first. Noise problems were cared for by an ingeniously mounted muffler once intended for a '42 Chevy. Aware that some Lime Rock locals were making as much noise as the cars themselves in protesting same, we made our post-prandial warm-up laps with the "thing" still in place. The green area begins at 4000 and that is more or less where we held it as we cruised the 1½ mile circuit. About the only noise we heard was the scream of gears from behind us whenever we backed off the gas. When Porsche went in for weight-saving, quietness went out of style!

Smoothness, too, though not for quite the same reason. As soon as we toured across the unpaved pit area, we noticed every little ridge and corrugation. The recently added Koni shock absorbers, though not adjusted to full tightness, make this car seem as rigid as a TC. Well, almost.

After several laps, the oil temperature needle had stirred itself off the peg enough to justify a little action. We'd already observed the extreme flatness of cornering and wondered why it seems to be that you feel the cornering side loads so much more strongly when a car does not roll much. It may be that the flatness so increases your security that you immediately commence more strenuous tactics. Or it may be that with firm bucket seats such as these, allied with tiltfree turns, it is possible to corner sharply without making extravagant efforts to brace oneself by arm or leg. Whichever the case, the Carrera seems to lead a strictly two-dimensional life, quite ignorant of the possibilities of motion about its longitudinal axis.

We did eventually press through these bends more rapidly, but even when we tried our utmost (see photograph) roll was minute. What gave us a clue that we were reaching some sort of limit was the feel of the steering. There seemed to be a gradual reduction in the force required to hold the wheels pointed in a given direction, although there was little if any actual change in the position of the wheel. This

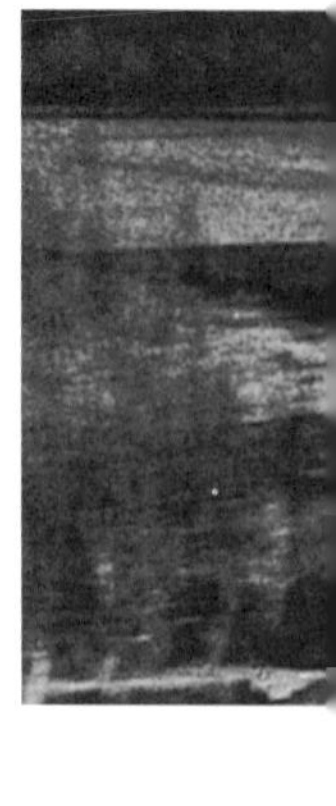

does not describe the difference in taking a corner two separate times but rather the experience involved in taking a single turn. It may be that we eased off a hair on the steering just as this reduction reached what we subconsciously classified as "minimum" but when you're that busy driving, it's hard to keep track of each and every action and its motivation.

Not to be neglected is the use of the throttle in cornering. As in any drifting machine, the engine's torque may be used to maintain or increase speed while aiding to hold the car into the turn. Backing off on the gas not only slows the Carrera, it starts the tail end moving out quite quickly. Briskly, you might even say. A light, sensitive foot is the thing to have and what it must be taught is to feather that accelerator pedal.

Once the oil was warm, we stopped to remove the muffler and ran from then on with that interesting Carrera option, the Spyder exhaust system. It certainly transformed the car as far as the bystanders were concerned, they could hear us all the way around the course. To the driver, the change not only provided considerably added push (before the engine felt to be dying at 7000 while it now took care to avoid exceeding the seven-five red line), it also proved to be a great morale-booster. As well, it became an aural tachometer and an indicator of wheelspin. If you corner too hard and get sideways, then it's all too easy to achieve the latter, if that can be considered any sort of achievement. That featherfoot technique is a must.

How to make a pushrod Porsche corner as well? A cynic might say trade it in on a Carrera, which is not a bad idea. For those of us who can't afford such a rash step, there are ways and means of modifying in the right direction. The first step is to have really firm shock absorbers. The Konis on Frank's GT certainly impressed us and so does the fact that they are now available in the USA. A good shock absorber shop with racing experience is a good bet, too.

The next step is de-cambering of the rear wheels. As readers of the VW or Porsche Workshop Manuals know, the rear torsion bars are splined on both inboard and outboard ends. The number of splines differs so that a vernierlike adjustment in increments of 50′ (5/6 of a degree) may be made in the setting of the trailing arm. This in turns controls the static camber angle of the rear wheel. The factory setting for Carreras puts both rear wheels at zero camber when the driver is in his seat. This is about as extreme as anyone should go for a car used on the roads; even then ground clearance can be a problem at times. Frank, however, has seen fit to further lower his suspension to get 1° of negative camber (wheels tilting inboard at the top). This step contributes towards more understeer or less oversteer, as does an increase in rear tire pressure. The Continental "Super Record" 5.90 x 15 tires were set at 28 and 32 psi, front and rear, incidentally.

The engine in the Wagenhofer Carrera is not the very newest. It still has its Bosch distributors mounted on the rear ends of the intake camshafts. Not only are they awkward to get at and adjust (they rotate in opposite directions, a point to remember when setting spark advance), they don't ease access to the eight 240T21 spark plugs. On the newer Carreras, the two distributors are mounted in a narrow Vee directly astern of the belt drive to the generator and cooling fan. This simplifies their adjustment and, to a small extent, the changing of spark plugs. Just reaching the furthest forward plug requires considerable dexterity, while changing all eight takes a good man at least forty minutes.

As Frank pointed out, this engine really requires talented maintenance by someone who's thoroughly interested. Fortunately, he has just such a man in his VW service department, one who has been to the Carrera school.

Red-lined by the factory at 7500 rpm, the Carrera's engine is fitted with two dual-choke Solexes. There's no progressive linkage here though, it's strictly one choke tube per cylinder. As tested, it carried 34 mm venturis, 102.5 main jets and 180 air correction jets. Before you rush out to re-jet your own Carrera, remember individual engines may be different enough to require slightly different jets, even on the same day at the same place. The only sign that road use had bothered this car was a slight misfire above 7000 that got worse as our testing went on. We were gla_ *we* didn't have to change the plugs.

The Carrera's clutch gets changed after every third race or so as it gets glazed quite quickly and then loses that positive bite needed when making fast shifts. This car has done seven races and about one-third of its 2100 miles have been covered on tracks, either racing or practicing. Other than clutches, the only difficulty was a broken second gear sycho-ring. Fortunately it happened at the SCCA National at Montgomery when all the necessary spare parts were in Frank's shop just a few miles away. A little night work and he was ready to go the next day. Go he did, too, getting a second to the unbeatable Pupilidy.

The joy of an all-indirect transmission is that individual gear sets may be changed without altering all of them. As well as having two ring and pinion choices (7/31 and 6/31, the 7/34 having been recently discontinued), there are three ratios available in each gear. Except for first gear, this Carrera has the lowest available ones; 3.09, 1.94, 1.23, and 0.96. With its 6/31 final drive and 5.90 x 15 tires, it reaches the 7500 red line at only 113 mph. It may not go fast, but it sure gets up there in a hurry.

Rumors for next year are quite firm that plain bearing cranks will be used in the Carrera Deluxes, not so sure is that there may be a 1600 cc alternative. There's even a whisper of a two liter version! But in Class F Production racing, it's as true today as it was in '53, the only thing that will beat a well-prepared, well-driven Porsche is another well-prepared, well-driven Porsche. *sfw*

It's not pretty but it does quiet the Spyder-like roar.

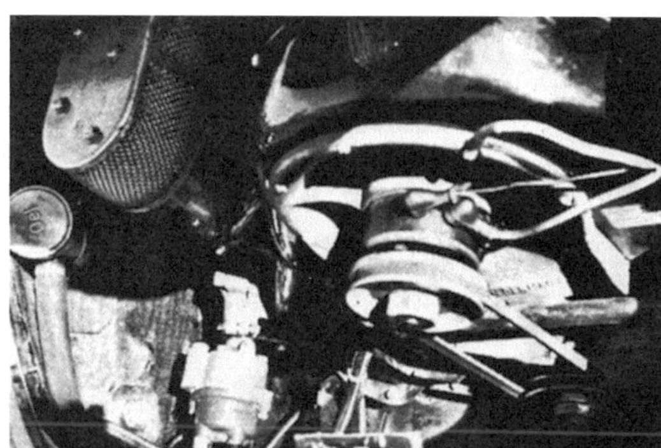

Spark plug changes are just plain difficult.

A prime example of flat cornering.

Left: Three of the first batch of Spyder RSK 1500s nearing completion at the Stuttgart factory. Right: A head fairing is incorporated in the detachable engine cover: bolt-on light alloy wheels are standard equipment

Porsche RSK 1500 GOES INTO PRODUCTION

A SEARCH through the race records reveals that few firms have been more consistently successful than Porsche in International events during the past few years. Last year in the Le Mans 24-hours race the works team, with RSK Type 718s, took third position in general classification with a car equipped with a 1,587 c.c. engine, and two 1½-litres of the same basic type finished fourth and fifth. One of the latter types, converted to a central seater, walked away from the opposition to win the formula 2 race at Rheims last June—and was beaten by McLaren's Cooper at the Nurburgring.

The Stuttgart firm have decided to go into production with cars similar to the 1958 works Spyder; when this became known in the U.S.A., eighty cars were ordered immediately. Seidel, Briggs Cunningham, McAffee and de Beaufort are among the well-known drivers who have already placed firm orders. Price in America is $8,000 (equivalent to approximately £2,850) which is considerably cheaper than the cost of Italian cars of similar class.

Available in 1½-litre form, the RSK has the horizontally opposed air-cooled engine with bore and stroke of 85 × 66mm (3.346 × 2.598in). Compression ratio is 9.8 to 1, there are two twin-choke Weber downdraught carburettors, the aluminium cylinder liners are "honeycomb" chrome-plated in the bore, and the power developed is 148 b.h.p. at 8,000 r.p.m., which represents 98.5 b.h.p. per litre. Mounted forward of the wheel centre line as in earlier Spyder models, the RSK engine has hemispherical combustion chambers with opposed valves operated by overhead camshafts. The drive to these is rather complex, by a series of bevels; there is a horizontal shaft to the lower (exhaust) camshaft and, from this, a short vertical bevel-shaft drive to the inlet camshaft.

Like the crankcase, the gear box casting is of magnesium-alloy, and is of the two-shaft or constant-layshaft type. There are four sets of ratios within the main casing—all equipped with the very efficient Porsche patented split spring ring type of synchromesh. Outside the main casing, an extra low gear for starting purposes and reverse gear are situated. A ZF type of limited slip differential is incorporated in the final drive. The gear box control lever is centrally mounted on the left-hand drive version.

Rear suspension is by a low point swing axle controlled fore and aft by a Watts linkage—very similar in arrangement to that used on the Mercedes types W 196 and 300 SLR. Combined coil springs and telescopic dampers are used for the suspension. At the front the Porsche type of trailing arms are retained but, compared with the firm's other production cars, they are much more widely spaced and use laminated torsion bars in K formation connected to each arm—an arrangement adopted, no doubt, to obtain a softer suspension with larger wheel movements. There is a telescopic damper at each side, and a transverse anti-roll bar connected to the lower arms. A tubular space type frame is used; it is arranged for left-hand drive, but can be converted to central steering and seating when required. The light alloy brake drums (with cast-iron liners) are 11in dia front and rear.

There are two front fuel tanks with a total capacity of 17.5 gallons and the dry weight is claimed to be 1,165lb (10.4cwt). With an overall top gear ratio of 3.45 to 1 the car has a claimed top speed of 155 m.p.h.

Left: Space type frame, showing the alternative mounting for the steering column to provide left-hand drive or central seating position. Below left: Coil springs surround the telescopic dampers for the low pivot-point, swing axle rear suspension. In the foreground is the rear upper arm of the Watts linkage used for fore and aft location of the axle. Below centre: The engine of the RSK is forward of the rear axle and the gearbox behind—a transposition compared with the units of the normal production cars. Below right: Front suspension, showing wide spacing of the trailing arms. Aluminium brake drums are turbo-finned, and there is a deflector plate for internal cooling of the shoes

Porsche for 1959

New RSK Type 718 Replaces 550 Spyder

PORSCHE have started production of the 1959 version of their sports-racing machine, Type 718, which will be known as the 1500 RSK 1959. Already the whole of the first series of about 20 cars, including several Formula 2 versions, has been sold. Price is 8,000 U.S. dollars, and buyers of the 156 m.p.h. machine include Jean Behra, de Beaufort, Art Bunker, Wolfgang Seidel and one of the Mexican Rodriguez brothers.

The engine is the usual horizontally opposed four-cylinder rear-mounted one, developing 165 b.h.p. at 8,000 r.p.m. A dry-sump lubrication system is employed, with an oil-cooler mounted at the front of the car. Two sparking plugs are used on each cylinder and two separate ignition systems, with twin coils, are used. Carburation is by two duplex down-draught Weber instruments.

Transmission details include a five-speed gearbox and the chassis is a welded tube space-frame. Front suspension is, naturally, independent, with two radius arms, torsion bars and duplex Koni telescopic shock-absorbers. Rear suspension is by low-pivot swing axles, wishbones and radius arms, duplex telescopic shock-absorbers and coil springs, used instead of torsion bars because of their lower weight. Dry weight of the car is 530 kgs., and it is though that the officially announced power output is the minimum, and that the actual output will be considerably higher.

GÜNTHER MOLTER.

Technical Data

Engine: Four cylinders, horizontally opposed, air cooled by radial fan. Bore 85 mm., stroke 66 mm. (1,498 c.c.). Compression ratio: 9.8:1. Power output: 165 b.h.p. at 8,000 r.p.m. Maximum torque: 14.9 mkg. at 6,300 r.p.m. Crankcase and single cylinders made from light alloy with hard chromium working surface on cylinder bores. Three roller and one ball-bearing to crankshaft. Four overhead camshafts, two coil springs to each valve. Transmission: Single dry-plate clutch, five-speed gearbox in unit with differential, with electron housing. Synchromesh on second, third, fourth and fifth speeds. Two duplex down-draught Weber carburetters.

Chassis and Suspension: Welded steel tube space-frame. Independent front suspension with two radius arms, torsion bars, duplex telescopic Koni shock-absorbers, and anti-roll stabilizer. Rear suspension with two low-pivot, swinging half-axles, wishbones and radius arms, duplex telescopic shock-absorbers and coil springs.

Steering: ZF-Ross, with damper.

Dimensions: Wheelbase 2,100 mm.; front track 1,290 mm.; rear track 1,250 mm.; length 3,600 mm.; width 1,510 mm.; height 880 mm.; dry weight 530 kgs. Estimated maximum speed: 156 m.p.h.

ABOVE: The welded-tube space-frame of the new car awaits finishing touches. BELOW: Front suspension incorporates two radius arms and torsion bars.

ABOVE: Rear end, showing differential housing and fan casing. BELOW: a racing engine—the well-tried flat-four, air-cooled layout.

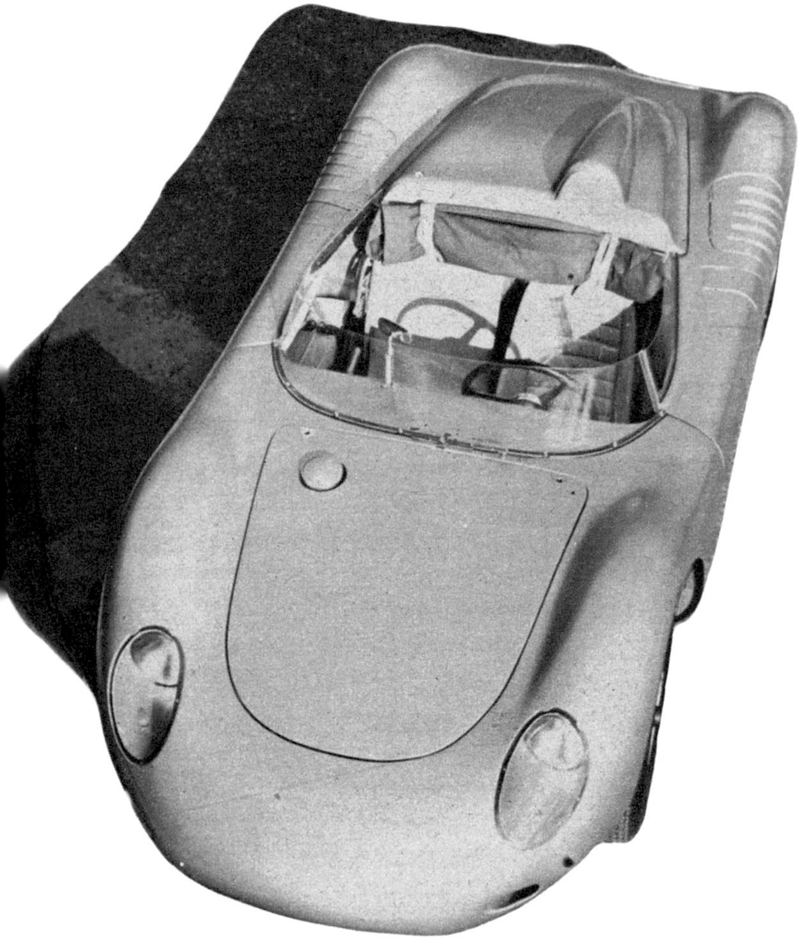

Only 20 to 25 Type 718s (RSK) will be made, but whole series is already sold out. Price: about $8000.

Low-pivot-point swing axle is located by wishbones and radius arms. Five-speed gearbox/differential is mounted to rear of engine. Crankcase, pistons and cylinders are light alloy, cylinder bores chromed.

POTENT

Porsche still hasn't gone the disc-brake route, apparently doesn't have to. Bi-metallic heavily-finned drums stop light cars deep in the corners.

Built-up crank has ball-bearing front main to take thrust, rollers center and rear. Rod bearings are rollers. Two dual-throat Webers supply fuel.

Bottom view shows finned cylinders and sump. Kingshafts running off the crank actuate lower cams, upper cams by connecting shafts. Two bearings hold short cams rigid. Individual cylinders are fitted with liners, domed 4-ring pistons scooped for valve clearance. Siamesed heads have hemispherical chambers, 2 plugs each.

Photo Story by Günther Molter

PORSCHE

Low-pivot effect is achieved by member (enclosing U-joint) pivoting from bottom of rear to give low roll center. Tank is overflow header for drive gears.

Space-frame, extremely rigid yet lightweight, is made from welded steel tubes. The two large upper and lower front crosstubes house torsion bars; rear has mounts for engine.

THE AIR-COOLED, four-cylinder engine for which Porsche is so famous.

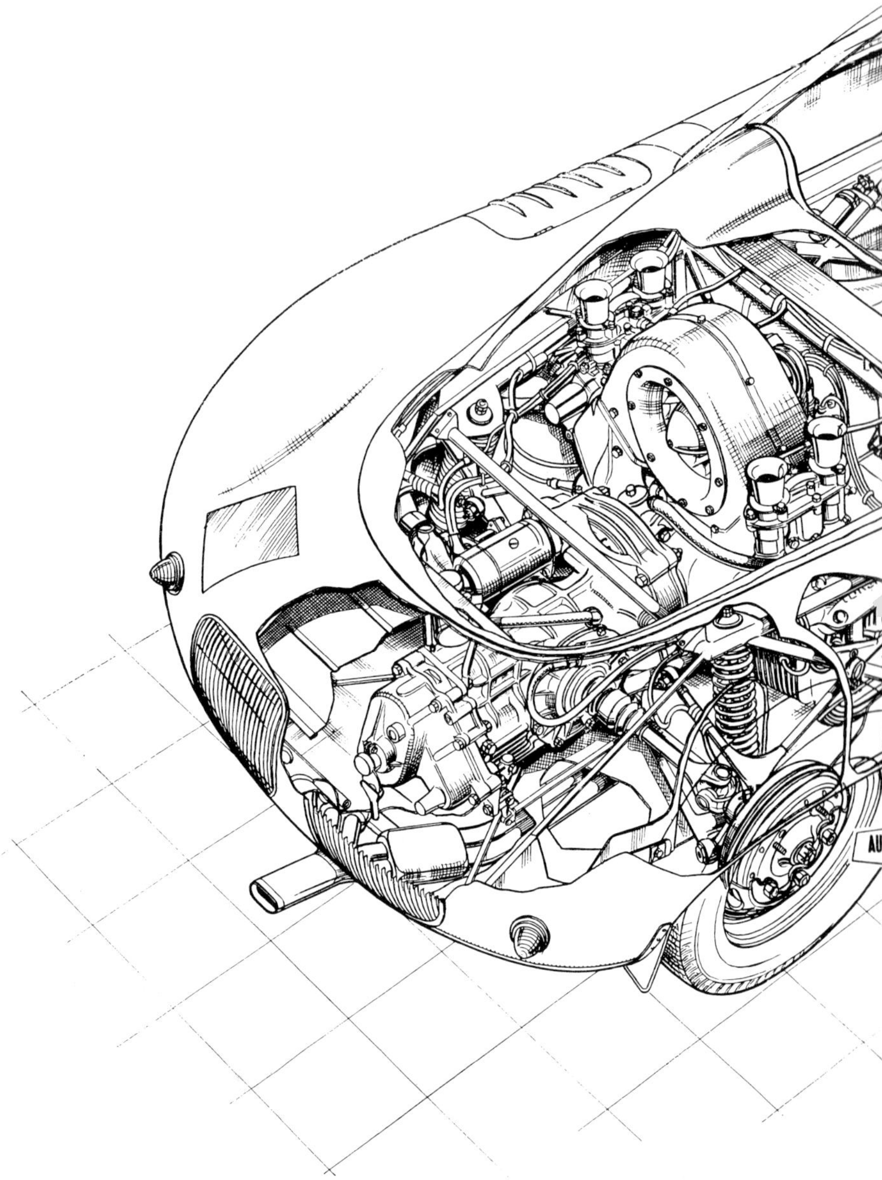

SUBJECT of Theo Page's fine drawing is the Type 718 Le Mans RSK Porsche "1500", which is also fitted with the 1,587 c.c. engine. In $1\frac{1}{2}$-litre form, the very efficient, air-cooled, horizontally opposed four-cylinder power-unit produces 165 b.h.p. at 8,000 r.p.m. on a 9.8 to 1 compression ratio. About 172 b.h.p. is claimed for the "1600" at the same engine speed. Maximum torque is 107.8 lb./ft. (14.9 mkg.) at 6,300 r.p.m. The normal RSK "Spider" gives 142 b.h.p. at 7,200 r.p.m., and 152 b.h.p. at 7,200 r.p.m. respectively.

The heavily finned light-alloy cylinder barrels are cast separately, with hard-chromium bores. Cylinder heads, also of light-alloy, are cast in pairs, with two valves per cylinder. Four separate overhead camshafts are employed, with a skew-drive to each cam-box, running through a tube integral with the head-casting. Dual helical springs are used.

Light-alloy is also used for the crank-case, which is heavily finned: dry sump lubrication is employed, the force-feed and scavenger pump assembly being driven by skew gears, at the base of the magneto drive. Two d/c Weber down-draught carburetters replace the Solex instruments formerly used.

The massive, counter-balanced crank-shaft is carried on three roller-bearings and one ball-bearing. Twin Bosch distributors each have separate coils (there are two sparking plugs per cylinder).

Transmission is taken via a single-plate dry clutch to a five-speed gearbox built in unit with the differential assembly, the whole contained in an "Elektron" casting. Bottom gear is non-synchromesh,

MASSIVE, counter-balanced crankshaft, carried on three roller-bearings and one ball-bearing.

The Porsche

A Theo Page

Description

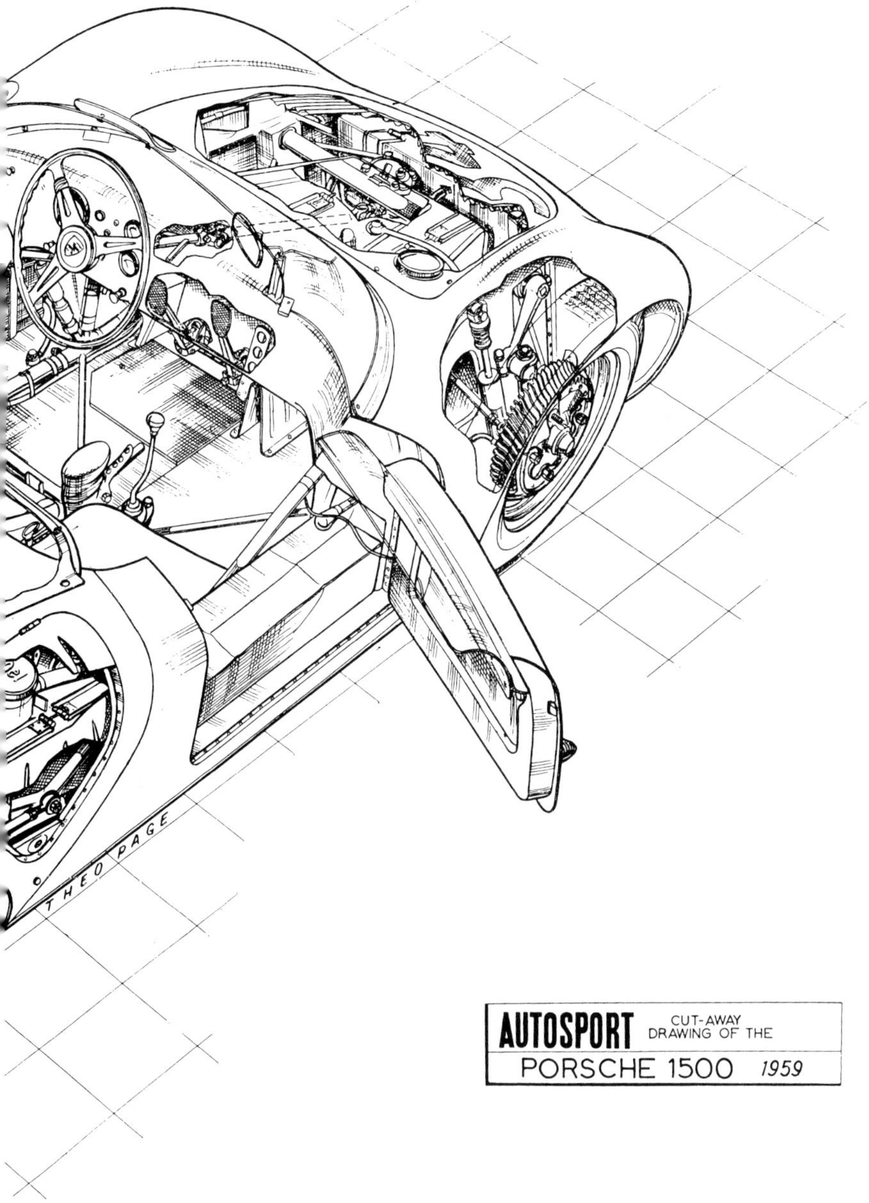

AUTOSPORT CUT-AWAY DRAWING OF THE
PORSCHE 1500 *1959*

pe 718 RSK

Drawing

or Grant

but the four upper ratios are on the well-known Porsche system.

Rear suspension is a development of the former low-pivot swingle-axle layout, but on the "718" widely spaced wishbones are employed in conjunction with helical springs, and radius rods. Koni hydraulic dampers are contained within the springs. At the front, torsion bars and radius arms are used, with trailing arms and an anti-roll stabilizing bar. Duplex dampers are fitted.

The frame is of welded-tube construction, carrying a light-alloy body. Fuel and oil tanks are carried in front, as is also an oil-cooler. Steering is ZF-Ross pattern. Front brake drums carry heavy, radial fins, but the rear are of the ribbed pattern. Wheels are bolted on to five studs.

REAR SUSPENSION on the Type 718 employs widely spaced wishbones in conjunction with helical springs and radius rods.

Wheelbase is 6 ft. 10.8 ins., front track, 4 ft. 2.8 ins., and rear, 4 ft. 1.2 ins. Dry weight is given as 1,169 lb.

For cooling purposes a turbo-fan is mounted directly above the engine, contained in a lightweight pressing. The main cooling draught enters the engine compartment via cooling ducts from the nose, and warm air is expelled from louvres on both rear wings and on the tail. Four separate exhaust pipes converge on a single tail pipe of large diameter, which protrudes well past the tail-cowling.

It is a remarkable tribute to development through racing, that this incredibly efficient "flat-four" engine had as its forbear the push-rod power-unit of the Volkswagen, designed during the 1930s by Dr. Porsche.

VIEW from below of the well-known "wasp-sting" exhaust—a familiar sight to many drivers of other marques!

TARGA CARRERA

by Jesse L. Alexander

► Violent, noisy, and with Spyder-like acceleration; this just about sums up the performance of the factory GT Carrera that we had the pleasure of driving recently in Sicily. The car, driven by Von Hanstein and Pucci, had just won its class in the 1959 Targa Florio by averaging 54 mph for the 1008 kilometers and finished in third place overall. Race numbers, straight through exhaust, a slight gash on the right door, and countless mashed bugs all over the front, attested to the ordeal to which this particular Carrera had just been subjected. Now it was patiently waiting for us to thrash it some more, so it could prove to us that it considers itself THE most reliable competition GT car on the road!

At Von Hanstein's urging, we slipped behind the wheel—down into the narrow bucket seat. Without the slightest hesitation, the Carrera kicked over on the first try. Behind us, the 130 hp 1500 Spyder engine with roller crank burst into life, its sound roaring at us through the thinly padded fire wall. We should point out that although this particular car was a factory special—having a 1½ liter Spyder engine and a special close-ratio gear box to fit Targa requirements, Porsche will make 1600 GT Carreras with similar performance for private owners—and with a plain bearing crankshaft. Backing out into the street from the small hotel in Cefalu where the Ferrari and Porsche teams were staying, a crowd of people gathered around to watch us take off up the road. With a shattering roar that drew other townsfolk to their windows, we rocketed away, the noise inside the passenger compartment making us think we were housed in with the engine.

In spite of the noise, it was amazing how untemperamental this Carrera was. It was most docile if one wanted to drive it that way, but we didn't. WHAM, went the throttle to the floor and the rev counter literally soared up to 7000 rpm in second gear. A snap shift into third and we were astounded at the immediate power. Admittedly, the lower ratios were extremely close, with only 700 revs separating them; also helping was the 6:31 final drive gear ratio, but the Spyder engine fitted with the dual throat Webers pro-

Above: At rest on the course the Targa Porsche—except for the numbers—looks like any other "P" wagon. Dog, donkeys, and farmers in the background show some of the problems confronting drivers on race day. Below: Enthusiastic Italians glad hand one of the team cars around a corner during the race. Needless to say, Sicilians love DRIVERS and fast cars.

Above: If any car can be called a dual purpose vehicle it is the Porsche Carrera. It can be driven in town—without developing extra muscles in your shifting arm. The other side of the coin is its ability on the race course. Its performance in competition has to be seen to be believed. Below: Very few 1500 cc cars can live with a Porsche on a tight course.

duced an incredibly clean and crisp acceleration. The gap between third and fourth was exactly 1700 rpm.

The brakes on the Carrera were, surprisingly enough, still quite potent. The shock absorbers were extremely stiff, however, and gave the passengers a jarring ride. Tenacious road-holding gives a driver full confidence in the car. Rear end breakaway as well as wheelspin was easily provoked by mashing the throttle. The gear box was, however, the high point of the machine. Perfect for hill climbs or sprint events, the close ratios do wonders for the performance of the Carrera, and we are of the opinion that this new Type 718 box now in production on all Porsches is the best yet built at Zuffenhausen. More direct and positive than before, the actual shifting effort seems to be markedly less—so great is the self-servo effect.

The engine in the particular car we drove was a 1½ liter Spyder engine comfortably putting out 130 horses. To correct for a peculiar vibration in the engine at certain revs, two distributors had been fitted but not hooked up to both upper camshafts. Another mod to this car was the installation of a 12-volt electrical system for a hotter and better spark at high revs.

Getting off the mark in the competition GT Carrera was much easier than in the de luxe model which we recently tested. This is caused in large part by the 6-31 ratio, but this modification is not recommended by the factory for normal use, due to overloads involved. Nevertheless, the car surged off the mark in smooth fashion, reaching an indicated 110 kph (68 mph) in exactly 11 seconds.

We spent over an hour behind the wheel of this special "Targa" Carrera, roaring about the Madonie mountains, through towns and villages, with people waving and shouting at us enthusiastically. It would not surprise us if future GT Carreras would be the *only* Carreras to be built, with power output closely related to the specs of this particular car. It was one of the most satisfying cars we've ever driven, for it GOES. Many thanks to Huschke von Hanstein for letting us try his special Sicilian bomb; Huschke, when do we get the RSK?

RIDE IN THE RSK

PHOTOGRAPHY: WEITMANN

by Richard von Frankenberg

► In at least one respect the Porsche Spyder is a rare chapter in the history of sports cars. It appeared on a race track for the first time in August of 1953. Yet after six years of development the Spyder – and especially its engine – looks virtually unchanged to the layman. This isn't uncommon in the field of passenger car design, where models are continued year after year with identical exteriors and engines that are or at least look the same – think of the Citroën 2 CV, the VW or many Opel types. But in the realm of racing sports cars, where performance is all that counts, it's almost unheard-of. Comfort, resale value, economy, amortizing the costs of tools – none of these viewpoints from passenger car construction has any meaning here. Thus all the firms that build racing sports cars change designs frequently and fundamentally to be able to "keep up" with the need for constantly improving performance.

Naturally the Porsche Spyder has also been altered and improved a great deal in years past, but often in ways which are invisible and which leave the basic design untouched. How and when did it originate, anyway?

It was clear to Porsche engineers in the Summer of 1952 that the production pushrod 1500 cc engine wouldn't be competitive much longer in international racing. For the Borgward sports cars were coming up strongly at home and the Osca – first with 1350 and later with 1500 cc engines – was walking away from Porsches abroad. As a result the first plans for a new engine were completed in late Summer of 1952. The bulk of the design work was handled by Dr. Fuhrmann, in the Porsche *Büro,* while significant stimuli also came from Ferry Porsche himself.

Named the Type 547 (every design carries its own number through the design offices), this engine was fitted with four overhead camshafts, which were driven by shafts and bevels. Like the normal Porsche engines, this was an air-cooled opposed four – thus two of the overhead cams ran along each side. From the beginning the engine had double igni-

Half-dressed RSK reveals new-type A-frame rear suspension. Chassis changes have been most responsible for improvement in lap times of Porsche sports-racers.

tion: two plugs per cylinder, two distributors and two coils. The two distributors were driven by extensions of the two upper (intake) camshafts. Today they're placed in a "V" and driven direct from the crankshaft. Unlike the normal Porsche engines, the 547 had a double-entry cooling blower, which means that the blower's scoops drew air from both sides. Two dual-downdraft carburetors brought the mixture in. In the six years that the motor's run, they've experimented back and forth between Weber and Solex carbs. At an original compression ratio of 9 to 1 the engine developed from 108 to 110 bhp (DIN) at about 650 rpm, the limit being about 7500.

If I'm not mistaken, the 547 engine ran for the first time on April 2, 1953 on the test stand in Zuffenhausen. This development was kept completely secret. In Spring of 1953 Porsche began to enter more factory teams in sporting events; until then the works had only sent light-alloy coupes to Le Mans, and had occasionally taken part in the Liege-Rome-Liege rally. But the Spyder with the four-cam engine wasn't yet available for the first races of the 1953 season. At the Eifel races it was Helm Glöckler who won for Porsche, but with a Spyder that was still powered by a hopped-up 1500 Super engine.

Then came Le Mans 1953. It was there that I first drove a Spyder which, if I remember correctly, wasn't yet called "Spyder". It was the Eifel Race car, though with an engine adjusted to normal gasoline to give just 78 bhp, fitted with a little coupe top which provided very poor vision and practically no ventilation – we won't even mention getting in and out. In any case the two cars (driven by Hans Herrmann/Helm Glöckler and Paul Frere and myself) were timed at 122 and 123 mph on the Mulsanne straight, speeds which are pretty respectable for 78 bhp and which show how good the car's streamlining was. Helm Glöckler and I drove one of these cars – open – in the Liege-Rome-Leige rally. The Belgian papers wrote: If these two Germans can stand 91 hours in this car and get back to Liege, they're the heroes of the rally. We failed to become heroes. Early in the third night, just as it began to rain, at Ponte di Legno in the Dolomites I leaned this Spyder rather ungently against a wall.

Next came the Grand Prize of Germany, which in 1953 was accompanied by a sports car race. By using alcohol the factory had extracted 98 bhp from the 1500 Super engine; that was the limit. With such an engine in the Spyder, Hans Herrmann turned a best lap of 11:02. During practice murmurs rippled through the paddock: Porsche has a new car here. Hans Herrmann made a few trial laps, and racing manager Huschke von Hanstein also took the wheel. The hood was never opened, and the Porsche technicians only shrugged the whole thing off. Ferry Porsche himself was there, for this was the first Spyder to be powered by the 547 engine.

In the Spring of 1954 at Hockenheim I drove this particular Spyder for the first time – it was called that then, after the American representatives had explained that this thing had to have a pretty name. Porsche had invited a number of the better German drivers to a day of trials, to enable them to choose a factory team. On this day it rained at Hockenheim. The three fastest on that day drove the works entries, together with Huschke von Hanstein, during the 1954 season. Hans Herrmann had the best time, then myself, then Helmut Polensky. Count von Trips first drove the Spyder in practice at Le Mans in 1955, and went very quickly from the beginning.

With the 1954 Spyder you could turn Nürburgring laps of between 10:50 and 10:55. In the hands of Grand Prize drivers, the RSK Spyder of today reaches the 9:50 mark and even a bit below, which represents an improvement of about 10 percent over 1954/55. In those days the very best time at the Hockenheim ring with the old Spyder was 2:33, while today we're closer to 2:21 and 2:22. This is an improvement of only about 7-8 percent. There's a difference because roadholding is much more important at the Nürburgring than at Hockenheim, and it is in this area that the car has improved enormously in the intervening years.

The first small step was the stiffening of the frame of the old "550 Spyder" with additional braces along the sides. Entered by the factory from the Spring of 1956, the RS Spyder had a completely new tubular space frame, while the RSK, tried experimentally in the Summer of 1957 and relied upon exclusively in 1958, had a completely different front end layout. Because the two upper frame tubes ran diagonally, forming a "K" structure, this was referred to as the "K axle". Porsche soon returned to a parallel tube arrangement, but the "RSK" designation was retained. To be a stickler, then, we have a car before us with an inaccurate type name. The RSK's rear suspension was fundamentally changed in 1958, with a switch from the traditional torsion bars to coil springs with concentric tubular shock absorbers (History and structure of the RSK or Type 718 are detailed in SCI, October, 1958, and the 550 Spyder in February, 1957 – Ed.).

Let me spend a little more time with this topic of roadholding. The old 550 Spyder was a most drastic oversteerer, an automobile that "came around" relatively early at the back if you were a shade too fast in a corner. It behaved very honestly in this range of breakaway, however, and was easy to catch and correct. In principle the RS had the same characteristics, with the exception that its tail end broke away quite a bit later. This motion showed up clearly, and here again you could correct relatively easily with opposite lock and application of throttle. Today's RSK is still describable as an oversteerer, whose steering is so close to neutral, however, that in certain situations (for example downhill on a low-friction road surface) it understeers. And since it gives no advance warning of breakaway, right up to the last instant, in my opinion the RSK is somewhat more difficult to drive than the RS. Or to put it another way it's no longer so carefree to drive, precisely because – and this may sound paradoxical at first – it has such fantastic roadholding, in the sense of gripping the surface.

On the other hand, on the basis of its power/weight ratio the RSK belongs to that class of cars that's steered about 70 percent with the steering wheel, and 30 percent or more with the accelerator – applying just the right amount of power and feeling for the point of incipient wheelspin at the back. With such power and light weight you can convert understeer to oversteer as required, by applying enough gas at the right time to spin the rear wheels at a faster rate than the front wheels are sliding out. Dry, the RSK carries about 7¼ pounds per horsepower. The engine outputs vary between 150 and 160 bhp (all DIN horses here), tending more toward 160 than 150. For the private cars the factory specifies 148 bhp (165 SAE!), but most of the customers' engines actually deliver somewhat more than 150. Very few factory engines produce more than 160 bhp, even when tuned for shorter races. The highest figures yet recorded on the dynamometer are 161 and 162.

In 1954 the gearbox was a normal four-speed unit. Only as a test, I drove a car with a five-speed box in the Tourist Trophy in September, 1955; this new box was standard on the Spyder from 1956 on. The RSK still uses it today, although a six-speed unit has now been developed for the Formula 2 car (see SCI, February, 1960–Ed.). Actually the RSK's five-speed transmission houses only four fully usable speeds, supplemented by a so-called starting gear. Possessing its own little auxiliary housing, the latter isn't synchronized and, like reverse gear, can only be engaged by passing through a lockout on the gate. Naturally the shift from the starting gear into second takes substantially longer than the normal change, because you have to pass over the lockout to get back into the main gearbox again.

Richard von Frankenberg holds it steady at 7500 while he has his picture taken by photographer Weitmann. New RSKs demand greater driving skill than earlier cars because they are faster.

Finned Type 718 Spyder had low pivot swing axle with coil springs in the rear. Interestingly, some drivers said car was better with fins, others thought it was more stable without the extra area, while still others couldn't tell either way.

Sans passenger, von Frankenberg tries really hard at the Hockenheim Ring. He has driven all the Porsche Spyder types from the 550 to the A-frame rear suspension cars being run now. These cars make a normal Porsche road coupe feel like an old, panting Volkswagen.

Between second and fifth speeds the gearbox is faster and easier to shift than on any passenger car that I know of. Even from third to fourth gear (which corresponds, in position, to second-to-third in a normal four-speed box), with no chance of a missed shift, you can whip through with astonishing speed, comparable only with the extremely short shifting times of a foot-shifted motorcycle gearbox. While the box of the normal Porsche makes a little noise if you rip right through from first to second, here you can move from second to third in the same way with no misgivings—presupposing that you can move the accelerator and clutch quickly enough, these being manual—or, more exactly, pedal—motions that can be learned. The transmission is a joy, also, because the shift lever "falls readily to hand" and because the shift pattern is much more compact than on a passenger car. These things are most intensely appreciated when—as during our acceleration tests—a hillclimb final drive ratio is installed which, going flat out, calls for engagement of fifth gear after four city blocks from a standing start.

What's the situation with rpm? Racing engineer Hild told me I could rev the engine to 8500 without undue concern. I know that the engine won't be damaged by brief excursions in the lower gears into the range just above, between 8500 and 9000 rpm. "Engine" in this case means valve gear, especially the valve fingers. Revs over 8500 are naturally not allowed in a long-distance race. A "spy" ensures that the mechanics can tell at once how much a driver has demanded from his engine: a second needle on the tachometer remains at the highest rpm figure reached.

This may be the most astonishing comment: the engine isn't at all inflexible. To be sure it doesn't accelerate strongly from low rpm, and when you step on the gas suddenly at 2000 it usually gulps before it bites off a chunk of fuel/air mixture. With some baffles screwed into the exhaust pipe, as a muffler, you can drive in the 1800-2500 rev range in town without roughness, as long as warming-up plugs are fitted.

The acceleration is fascinating, much more for a passenger than for the driver, who always has the steering wheel to hold on to. The passenger is pressed back against the seat so firmly that he's unable to move for a few moments. But allow me to express this in figures. We chose the classic European acceleration distance, the kilometer (5/8 of a mile) from a standing start. I made the trials at Hockenheim, on the start-finish stretch with its slight grade of about 0.7 to 0.9 percent. On the first run, slightly uphill, I took the RSK to 8400 in the gears and up to 8500 in fifth. The gearing was just a shade low for the kilometer, so I couldn't quite hold full throttle over the last 300 yards. This took 25.8 seconds. Within the last year Erni Vogel made a record attempt in Austria with a Spyder and reached 26.01, which was a new Austrian record. The international record in the 1500 cc class has stood for a long time at 24.94 seconds (average of two runs), and was set in 1937 by a blown 1½ liter Maserati racing car on the *autostrade* near Florence. Naturally the postwar racing cars of the same displacement—the Alfetta, for example—were faster, but no one ever took the trouble to put them to the test.

On the return run I turned the engine to 8700 in the gears and to 9000 in top which resulted (admittedly down the tiny grade) in a time of 24.5 seconds, 4/10 of a second under the record. The average, however, was 25.15 seconds. To be honest, I had previously made a practice start, then came the two violent starts for the standing kilometer—and then it was all over for the clutch. In accordance with one of the rules of the game within the Porsche racing team, applicable whenever one of us damages something in the powerplant, I gave the racing mechanics a case of beer.

We didn't make any top speed runs, but we do have one time taken at Avus in 1958. Only on a course like Avus or Rheims, with long, level straights and approaches, is it possible to get decent top speed indications. At Avus Behra's car, an RSK Spyder with 160 bhp, reached exactly 160 mph.

The pedals are naturally so arranged that you can "heel-and-toe" between brake and accelerator—which means simultaneous braking and revving for downshifts. What most race watchers and laymen don't believe: Although the transmission's synchronizers are near perfection, we never fail to double-clutch on downshifts in races as well as on test runs. You mustn't forget that a gearbox that has to carry 8000 or 8500 rpm is most grateful for a little assistance from double-clutching!

While shifting itself seems to go much more easily than in a passenger car, the pedal pressures required are substantially greater. A man who drives a Spyder for the first time and, as is suitable, shifts frequently and also brakes often from high speeds will find that his feet are weary after a very short time. The brakes are fascinating. Even at very high speeds they can produce deceleration between 0.9 and 1.0 g, and under favorable circumstances (high-friction road surface) even a hair more. If, at about 140 mph, you begin braking hard just 200 yards from a hairpin turn, you'll always make it around the turn with safety. If you experience this as a passenger in a Spyder, without ever having tried such brakes, it seems like a wonder of physics. When braking hard on a surface that has shallow ripples—ripples which you often can't see with an unpracticed eye—it's necessary to make gentle sawing movements at the wheel to keep a single brake from locking up. Generally they never lock up completely, but only for a fraction of a second which leaves a black stripe on the asphalt. But at high speeds that's enough to throw the car off its course.

You get through curves quickly with the RSK this way: Just at the point where you begin to turn into the corner you put your foot back on the accelerator and try to feed an amount of gas such that a) the back wheels won't spin, which would put the car sideways in an instant, and that b) the rear wheels have just enough slip to put the car into a gentle drift. If this is done right you won't have to apply much more steering lock, since the entire car will be placed at a tangent to the inner edge of the curve. You mustn't apply full throttle too soon coming out of the bend, for the RSK will reply with breakaway at the back which costs valuable fractions of a second —even though it's quickly recoverable by steering into the slide. When emerging from corners it's also imperative to keep an eye on the tachometer to avoid exceeding the prescribed rev limit. Since the RSK has no speedometer, a racing driver himself can never tell just how fast he's taking a certain corner. He can only tell with what revs and in what gear he leaves the curve; these are enough to decide how fast a lap he turns.

You must wear a pair of racing goggles. When I took the photographer, Julius Weitmann, for a nearly flat-out lap of Hockenheim, he elected to wear no goggles to be able to operate the camera more easily. Afterwards he told me that above 120 he could only open his eyes slightly, and that with great difficulty, since the air rushes back over the low windshield right at eye level.

If you step back into your 1600 Super Porsche after an afternoon of driving an RSK, you'll think someone's installed a Volkswagen engine while you were away. When you try the brakes you'll think they're broken—not because of the somewhat lower deceleration, but because the production car's pedals work so much more easily than the RSK's. Also, of course, the virtual silence of the Super (compared to the RSK) heightens the feeling that you're driving a very tame automobile.

With the RSK, or with any good 1500 cc sports/racing car, better times can be put up today on courses like the Nürburgring than 2½-liter Grand Prize cars could reach in 1954. Since the sports car's top speed must be lower, improved roadholding made the difference. This is a fine example of the effectiveness of engineering and of genuine technical advance—to the eventual benefit of the production car.

—R V F

letter from europe

PORSCHE, 1960

Two new ones—RS60 racer, Abarth Carrera GT

RS60 series production sports-racer shows Porsche has given up worming around FIA's windscreen rule. Frame is wider than RSK's.

FOR 1960 Porsche has built a small series of sports-racing two-seaters in an improved version for 12 internationally qualified private owners. Seven go to the U.S.A., two to Canada, one to South Africa and two to Europe. Price at the works in Stuttgart is £4500.

The car, called the RS60, differs in several respects from the 1959 type 718. Bodywork has the F.I.A's new regulation windscreen 9.8 in. high. Space has been found for a small luggage compartment behind the engine and above the gearbox at the back.

Front suspension has the usual twin trailing arms attached to transverse torsion bars, but it is re-designed to take Koni 85/75 double-acting telescopic dampers. At the rear is the new double wishbone independent suspension seen at the end of 1959.

Drive shafts are double-jointed wishbones, semi-trailing. Suspension itself is by coil springs enclosing Koni 100/85 telescopic dampers.

Wheelbase is now 86.6 in. against 82.7 in. last year. Wheel size is reduced from 16 to 15 in. Frame is wider and car is now 145 in. long, 59.4 in. wide. Height to top of windshield is 38.5 in. and ground clearance is 5.9 in. Front track is 50.8 in., rear 49.2 in.

Engine is the well-known Carrera unit, with a specially reinforced crankshaft. It is available at 1498 cc (85 x 66 mm) giving 147 bhp installed, or at 1587 cc (87.5 x 66 mm) giving 157 bhp. Maximum rpm is 8000, top speed 156 mph.

Drum brakes are still used front and rear, with turbo cooling fins. Drum diameter is 11 in. Total brake lining area at front is 86.5 sq in., rear 57.6 sq in. Car is a little heavier than last year's

Abarth's GT Carrera is a hyper-streamline job, features adjustable intake scoop.

at (dry weight) 1220 lb. Fuel capacity in the nose tank is 17.6 gallons, oil capacity in side tank is 1.76 gallons.

The Formula Two car on which the new wishbone rear suspension was first seen gets only minor changes. They include a slightly slimmer body. The car was due to be driven for the first time by Stirling Moss at the Syracuse G.P. on March 19.

Abarth has produced a new super-lightweight body for the Porsche Carrera GT. He is busy making a series of 20. Special care has been taken to keep the glass as near as possible flush with the surface of the body. Even the windscreen wiper feeds are covered by small fairings formed in the nose panel!

The tail has an adjustable air intake scoop, and there is a ducted oil cooler below the nose.

Weight is 1750 lb., around 95 per cent of the weight of the standard model. The Abarth car should therefore be eligible for homologation as a GT racer. Maximum speed is claimed to be 135 mph.

Price at works, about £1700.

Porsche's Formula Two single-seater, new last season, gets minor changes for 1960. Among them is a narrower, generally cleaner body. Stirling Moss will drive a Porsche in Formula Two races this season, but his car will not be works-maintained. Instead Porsche is turning it over to the care of British whisky millionaire R. R. C. Walker's racing stable. Rob Walker, of course, also owns the F-1 Cooper with which Moss will make his all-out bid for world championship honours.

—*Gordon Wilkins*

No regulation luggage space here! Boot is at back instead, behind 157hp engine.

THREE NEW PORSCHES

IN PREPARATION for the 1960 racing season, Porsche has developed two new cars. The new Formula II car is basically the same as the 1959 car, with only minor changes, the most noticeable of which is a slimmer, more attractive body. The engine and running gear is practically identical with that of the RS-60, but carries a single-seat body and 6-speed synchromesh transmission. Moss, Bonnier and Graham Hill are scheduled to drive the F-II Porsches during 1960.

The RS-60 *Rennsportwagen* has undergone considerable change from the RSK type-718. Wheels are now 15-in. in place of the former 16's, carrying 5.50 x 15 front tires and 5.90 x 15 rear tires. The wheelbase was increased from 82.6 to 86.6 in. and the tread is 50.7 front and 49.2 rear, overall length is 145.6 in. and overall width is 59.4 in. Height, to the top of the FIA-required 9.8-in. windscreen, is 38.6 in.

Two engines are available for the RS-60: 1498-cc (85 x 66 mm bore and stroke) with 150 bhp, and 1587-cc (87.5 x 66 mm) with 160 bhp. Maximum rpm is given as 8000 and Porsche claims 156 mph.

Front suspension is still trailing links with transverse torsion bars and at the rear the double wishbones introduced at the end of 1959 are used with coil springs. Koni telescopic dampers are used all around and drum brakes have been retained (11 in. diameter). Weight has been increased and the cars now tip the scales at 1110 lb dry.

Only 12 RS-60 Porsches are scheduled for production and will go to qualified private drivers only: 2 in Europe, 2 in Canada, one in South Africa and 7 in the U.S. The price is reported to be $9000 per car, delivered at the factory.

The RS-60 Rennsportwagen is longer and heavier, but faster.

Hans Herrmann tests the new Formula II Porsche . . .

. . . which lacks paint but is otherwise ready to go.

It also comes equipped with roll bar and luggage space.

BY KURT WÖRNER

PHOTOS–LEICA STUDIO–WÖRNER

And Abarth gets into the act with this sleek, handsome Porsche Carrera. We'll have more on this later.

► At Porsche the new racing coupe is known as the "Abarth car" because its body was designed in Turin by Carlo Abarth's engineering staff and Scaglione. Solid ties between Porsche and Abarth have a historical basis and Ferry Porsche wanted very much to let Carlo have a chance to design a new racing Carrera to replace the normal GT. After considerable negotiation, a series of 20 356B chassis was sent off to Turin last fall.

As soon as car number one got back to Stuttgart with its Italian clothes, several shortcomings were immediately discovered. First of all, the interior had obviously been designed for an Italian, for virtually no one at the Porsche factory could sit in it. This severe lack of head room was accompanied by an almost complete absence of any steering lock. Body tools and snips brutally altered the shape around the wheels and a solution to the headroom problem had to be found before the car could be raceworthy.

DRIVER'S REPORT:

PORSCHE CARRERA GTL

by Jesse Alexander

Porsche wasted no time in putting its new car into competition, and its first race was the Targa Florio. A second GTL was raced in the 1000 Kilometer race on the Nürburgring, fitted with experimental disc brakes, then on to Le Mans with production brakes installed. On every occasion the Abarth Carrera exhibited impressive acceleration and a remarkably high top speed (timed on the Le Mans straight at 137 mph) leaving no doubt that its aerodynamics are superior to the production car. This, plus the fact that Abarth had succeeded in paring at least 330 pounds off the catalog Porsche (down to 1765 pounds), has resulted in real race-car performance, exactly what Porsche was after.

Sheer light weight was the prime requisite. The interior is stark, with pull-strap windows and plexiglas used where possible, no attempt being made to reduce noise or make the car luxurious. It is designed for weekend racers who, for approximately $6000, can have a Carrera giving impressive performance with an amazingly docile and flexible engine that appears to be as happy pulling at 2000 rpm as it is at 7400.

Huschke von Hanstein lent me the Le Mans GTL shortly after its return from the Sarthe circuit. Nothing had been done to the engine and everything except the hand brake was in working order. Herbert Linge accompanied me to the car, warning of its outstanding ability to ship water and, in his wonderfully accented English, relating the misery of sitting in a soaking wet seat for the better part of 24 hours. He also had to contend with several inches of water filling the bottom of the Carrera.

Pointing out the position of wiper and headlight switches and with a word of caution not to exceed the indicated 7400 rpm limit, Herbert left me to my pleasure. I turned on the silver coupe, still with its huge black Le Mans numbers painted on the sides and rear. On cracked throttle the Carrera fired immediately. Expecting a fierce clutch action, I was astonished to find that it was as tractable as a production Porsche. Stiffly sprung, the GTL still doesn't seem to have as harsh a ride as certain Super 90 Porsches on uneven surfaces, apparently due to a new type of Koni shock absorber. In a minute I was on the Autobahn, accelerating away like a rocket, passing slower traffic as if it were standing still.

A fast run on the Autobahn in a sports-racing car can be an amusing experience; even the police turn and look with envy as you roar by at 110 mph. Up through the gears using 7400 rpm, the Abarth Carrera arrived at a peak engine speed of 6900 rpm in top gear, the speedo hovering about the 130 mph mark and the rev counter trying ever so hard to knock off 7000 in top. Engine noise is brutal at high speeds and if a window is lowered the noise and turbulence is violent. This four-wheeled projectile is pointed rather than steered, and can be classified as a Spyder with a roof. A few of the latest GS four-cam plain-bearing engines are giving 135 bhp, most of the horses coming in with a punch at 5500 rpm.

A superb driving position is one of the finest attributes of the Abarth car. Getting behind the wheel is another story, the occupants being forced to double over almost in half and ease in slowly, so low is the roof line. But behind the wheel, the driver has exactly what he wants within easy reach or sight. A revised instrument layout places a huge revolution counter, complete with red tell-tale, and other pertinent instruments directly behind the wheel and slightly higher than on the production Porsche. An oil pressure gauge is positioned on the floor directly behind the gear shift lever, and behind the seats a solid roll bar comes up out of the floor from both sides. The distance between the top of the driver's head and the roof is minimal yet adequate, but anybody over six feet would have a real problem.

Off the Autobahn, I drove to Stuttgart's Solitude circuit, a beautiful old-Watkins-Glen type of course that rivals the Nürburgring. The Carrera was fitted with German Dunlop "SP" belted tires which contributed greatly to its superb roadholding. Viceless and well-balanced are the two terms best describing current production Porsche handling, and they apply equally to the Abarth Carrera – with the single warning that breakaway is reached sooner and more easily with the high power-to-weight ratio.

The Porsche sales department reports that the entire series of 20 Abarths is already sold. The problem at present is to improve its detail finish, which is not at all up to normal Porsche standards. Delivery is also way behind. It's a safe bet that the line of the Abarth car could influence future Porsche styling, and taking a long guess I can easily visualize a 1962 Porsche looking very much like this Italian/German effort. A slightly longer wheelbase and revised entry and exit would make it more comfortable and give the occupants more than just the present acceptable minimum of space. All kinds of interesting things can be expected from Porsche in the next few years and the Abarth-bodied GTL Carrera should be regarded not only as a racing GT car but as a stepping stone to new and exciting things to come. *—JLA*

After watching the GTL for 24 hours at Le Mans, SCI's European Editor finally gets his hands on it. He says police watch in envy as you roar by at 110.

Despite pretzel twist needed to get in or out, the GTL is comfortable to drive — when it's dry. In rain it ships water that won't leak out.

Large tach with tell-tale dominates instrument cluster. It's flanked by oil temp/fuel gauge and 250 kph speedo. Oil pressure gauge is on floor.

Un-Porsche-like rear has adjustable airscoop. Interior hardware, steering wheel, gear lever are standard items.

A Limited Edition

Abarth Porsche Carrera illustrates broad versatility of overseas designers

THE "UNCHALLENGED OVERSEAS SELECTION" we mentioned at the beginning of this section is possible for many reasons. One reason is illustrated here—the installation of a distinctive custom body on an existing chassis by one of Europe's many limited-production coachbuilders. Germany's Porsche is the subject and Abarth of Italy is the firm that has transformed the Carrera GT chassis into an exciting coupe that is as fast as it looks.

Plans are to build 20 cars in this truly limited series. While this makes them rare, it is most certainly not a car that would sell in large numbers even if Abarth could produce more. Utility in this type of automobile is restricted to competition or high-speed fun driving; it is an expensive (no price available) plaything. The point is that many of the world's most famous and talked-about cars—cars that help make up the wide selection from abroad—are similarly rare; yet they are produced and can be purchased if money and desire are both present in equally generous quantities.

Abarth's body is super lightweight. Special care has been taken to improve streamlining; for example, even the windshield wiper feeds are covered by a small fairing formed in the nose panel, and the glass itself is as nearly flush with the body as possible.

The high-performance, four-overhead-cam engine, essentially a de-tuned Spyder powerplant, develops 115 hp (DIN rating) from 1600cc. This is one very fast automobile.

PORSCHE "PROTOTYPE"

NOT CONTENT with the performance of the 4-cyl Carreras, Porsche has now produced a "prototype" coupe based on its flat-8 GP single-seater. For use in this "GT" car, the flat 8 is opened up to 2 liters and, in its present form, is fed by 4 dual-throat downdraft Weber carburetors. Output is 210 bhp at 8400 rpm.

The chassis incorporates most of the features from the GP layout but is, of course, somewhat wider. An interesting item is the use of transfer gears at the inner ends of the axle shafts to bring the engine and transmission forward, presumably for the purpose of improving the car's balance.

The bodywork is, we hope, efficient. Certainly, it is most unusual and would probably create quite a stir in traffic.

PHOTOS BY EBERHARD SEIFERT

Putting the prototype 2-litre Carrera through its paces.

An earlier Speedster, very popular in the U.S.

Maglioli's Porsche in the 1959 Targa Florio.

"Some say that only Volkswagen's finish is better!"

PORSCHE

Some Heckmotor history related by

WHICH was the first-ever GT car? The Lancia Aurelia coupé, often quoted, did not appear on the scene until 1951. At the 1949 Geneva Automobile Show the Porsche 356 was introduced to the world. Unless more knowledgeable readers of *The Motor* are able to dredge up any other candidates, I stand fast and cite Ferry Porsche's two seater as the instigator of a post-war trend that is far from over. Enduring for over 13 years now, the Porsche 356 is unique in that its basic dimensions are still the same as they were when the car made its début. The entire concept of the automobile was so outstanding that there has been no need for drastic change.

The Volkswagen based, rear-mounted, air-cooled four-cylinder engine has slowly been developed and refined over the years, until this pushrod 1½-litre is now at the height of its development, and Porsche are expected to replace it in the near future. Periodic, meaningful modifications have been made over the whole car, while its original "perfect" shape has remained unaltered. The Porsche boasts one of the highest standards of finish to be found on any car, regardless of price.

I first saw a Porsche "in the flesh" in California during the summer of 1952. There were only a few in the U.S.A. at that time but already the car was being raced in Europe (winning the 1,100 c.c. class at Le Mans in 1951) and attracting international interest. I have never been so enamoured with a vehicle in my whole life as I was with that red coupé that so fascinated me ten years ago. Since then I have owned three Porsches: a 1500 drophead, a 1600 coupé, and a Super 90 coupé. When you own a Porsche you feel as if you belong to a family, and affection for the automobile grows the longer you own it. As a means of personal transport, the car is ideal for one person; for two people and luggage one has to go easy on the number of bags, but fitted luggage, which can be supplied by the factory, all but eliminates this problem.

If there is one factor that sells a Porsche, it is the fine quality of finish which hits the prospect full in the face the moment he sits in the car. In spite of production increased to 35 cars a day the Stuttgart-Zuffenhausen factory has been able to maintain its standard of hand-finished quality. Some say that only Volkswagen's finish is better!

Move to Stuttgart

The 356 was drawn up in 1947 by Ferry Porsche and chief designer Karl Rabe while Ferry's father was still in a French prison. The initial series of 50 cars were all made by hand, had light alloy bodies and used an 1100 VW engine. The factory was moved from Gmund, Austria, to Stuttgart after the series was built, and there the Professor took up again where he had left off at the close of the war. From 1,086 c.c., the displacement of the model 356 Porsche was increased to 1,300 c.c., then to

…e Mans prototype. Some exhaust pipe!

Mechanics fitting a rear suspension. Note the finned brake drums.

…935: Dr. Ferdinand Porsche concentrates on an Auto Union.

Von Frankenberg and a Spyder on ice.

By Jesse Alexander

…well-known American journalist who has owned three himself

1,500 c.c. in 1952 when the 1500S with its famous Hirth roller-bearing crank was introduced. All three displacement sizes were offered for sale until the 1600 engine was introduced in 1955, either in "normal" form or as a "Super". Leopold Schmid, one of Porsche's most brilliant designers, saw his fully synchronized four-speed block ring transmission fitted in the car as early as 1952, replacing the VW crash box that was the Porsche's original equipment. As the car received more and more performance, chassis modifications to improve its braking and handling kept pace. Porsche have remained true to drum brakes right down through the years and only in mid-summer, 1962, did the factory switch over to disc brakes of their own design for use on the 130 b.h.p., 125 m.p.h. Carrera 2.

The "tight little coupé" has been supplemented by drophead and hardtop models, the bodies built by Reutter and Karmann. Several years ago, Drauz in Heilbronn were turning out a so-called "Speedster" body for the 356 chassis which proved extremely popular in the U.S.A., especially in California where the climate matched the car's sparse weather equipment. Dan Gurney, currently Porsche's top Formula driver, began his career in a Speedster.

Racing has been part of the Porsche story right from beginning. Professor Ferdinand Porsche designed the pre-war Auto Union cars driven by Stuck, Varzi and Rosemeyer. The post-war Porsche racing shop has always been a part of the design and development office and several of the late professor's collaborators still contribute to Porsche racing car design. The mid-1950s saw the *Rennsport* (sports-racing) Spyders at their zenith; the 1959 overall win in the Targa Florio is considered by many to have been their greatest post-war competition success. The Nürburgring, Le Mans, Liège-Rome-Liège, have all been won by the Zuffenhausen cars at one time or another and virtually every top-ranking driver of the last decade has been behind the wheel of a Porsche. Racing manager Huschke von Hanstein travels all over the world with his team; I first met him in 1953 in Mexico when he was looking after Karl Kling's 1500S pushrod Spyder in the Carrera Pan-Americana.

Carrera Gets its Name

The word "Carrera" has been associated with Porsche since the Mexican road race was run for the last time in 1954, when the German cars were first and second in their class and third overall, for Carrera was the name given to the tamed GT version of the four-cam Spyder engine designed by Dr. Fuhrmann of Porsche in 1952. The Carrera appeared at the 1959 Frankfurt Motor Show and in its 1.5-litre state was rated at 100 b.h.p. The car had a most satisfying performance, albeit at a handsome price.

Continued overleaf

When finish isn't the prime factor.

Porsche only change to improve. Enlarged rear window on Super 50.

Porsche jeep prototype, never produced.

Porsche make helicopters too.

1962 Targa Florio: Maglioli (Porsche RSK) squeezes through on the inside.

PORSCHE

Maintenance on these cars turned out to be tricky as well as expensive and there were so many headaches, both in production and in the servicing of them that this touring Carrera was quietly dropped from production. Private owners continued to race Carreras, however, and factory tuned engines were recently developing up to 130 h.p. Carlo Abarth got into the act in 1960 when he was commissioned by Ferry Porsche to have an Italian body fitted to a limited number of Carrera chassis. Strictly for competition, all of these Abarth-Carreras were sold almost before delivery. Recently the Carrera 2, a highly refined 2-litre touring Carrera, has been added to the Porsche line.

An RSK Spyder with central driving position appeared at Rheims in 1948, heralding the participation of Porsche in Formula racing. Jean Behra drove this car to victory in the Formula II race, beating Peter Collins' V-6 Ferrari. Exactly a year later, the first proper Formula Porsche appeared at Monte-Carlo, driven by vonTrips, the chassis based on RSK principles and the now familiar four-cylinder Spyder engine in the usual position at the rear. Trips distinguished himself that weekend by crashing the Porsche on the second lap of the race but this misfortune in no way diminished Ferry Porsche's growing interest in 1.5-litre Formula racing; a month later Jo Bonnier was driving the rebuilt machine at Rheims, finishing third behind Hans Hermann in Behra's Formula 2 Porsche.

But the end of the four-cylinder 4-cam engine was in sight, at least for racing, and Porsche hoped to have a new eight-cylinder to replace it when the 1.5-litre Formula I came into force in 1961. Progress was slow, however, and respectable horse-power figures proved elusive. A whole year elapsed before the engine was ready and two 2-litre eight-cylinders were brought to Sicily for the 1962 Targa Florio. The engines proved reliable, but brake problems hampered both cars and were their eventual undoing. Meanwhile considerable testing had been going on with a Formula I chassis at Hockenheim and on the Nürburgring and the first Grande Epreuve at Zandvoort saw both Gurney and Bonnier driving brand new cars.

Formula 1 Flat 8

The race was won handsomely by Graham Hill in his B.R.M. but not before Dan Gurney had been heard from in the opening laps. With the engine again showing great promise, chassis

is the car the works will be contesting long-distance G.T. races with. So take a close look at Porsche's 904—you'll be seeing a lot of it next year!

904 G.T.

ent Trend of Rear-Engined beatable in 2-litre Class

universal joints, which have two pins side by side in what is really a short shackle, thus deleting the usual splined slip-joints. The pivots of the rear wishbones are arranged to give a trailing angle and in front there is a forward inclination.

The body is of glass-fibre, with a steeply inclined curved front screen and a flat rear window recessed right into the body. Behind the rear window, a slatted air intake stretches across the tail, and there are intakes for cooling the rear brakes on either side. Sliding windows are fitted in the doors. A single windscreen wiper blade, with double arms, remains vertical as it slides across the glass.

The wheelbase is 7 ft. 6½ ins., track 4 ft. 3¾ ins., and overall length 13 ft. 5 ins. The body shape is highly advantageous, but the greatest advance is in weight saving, the 904 being 3 cwt. lighter than the previous Porsche G.T. model.

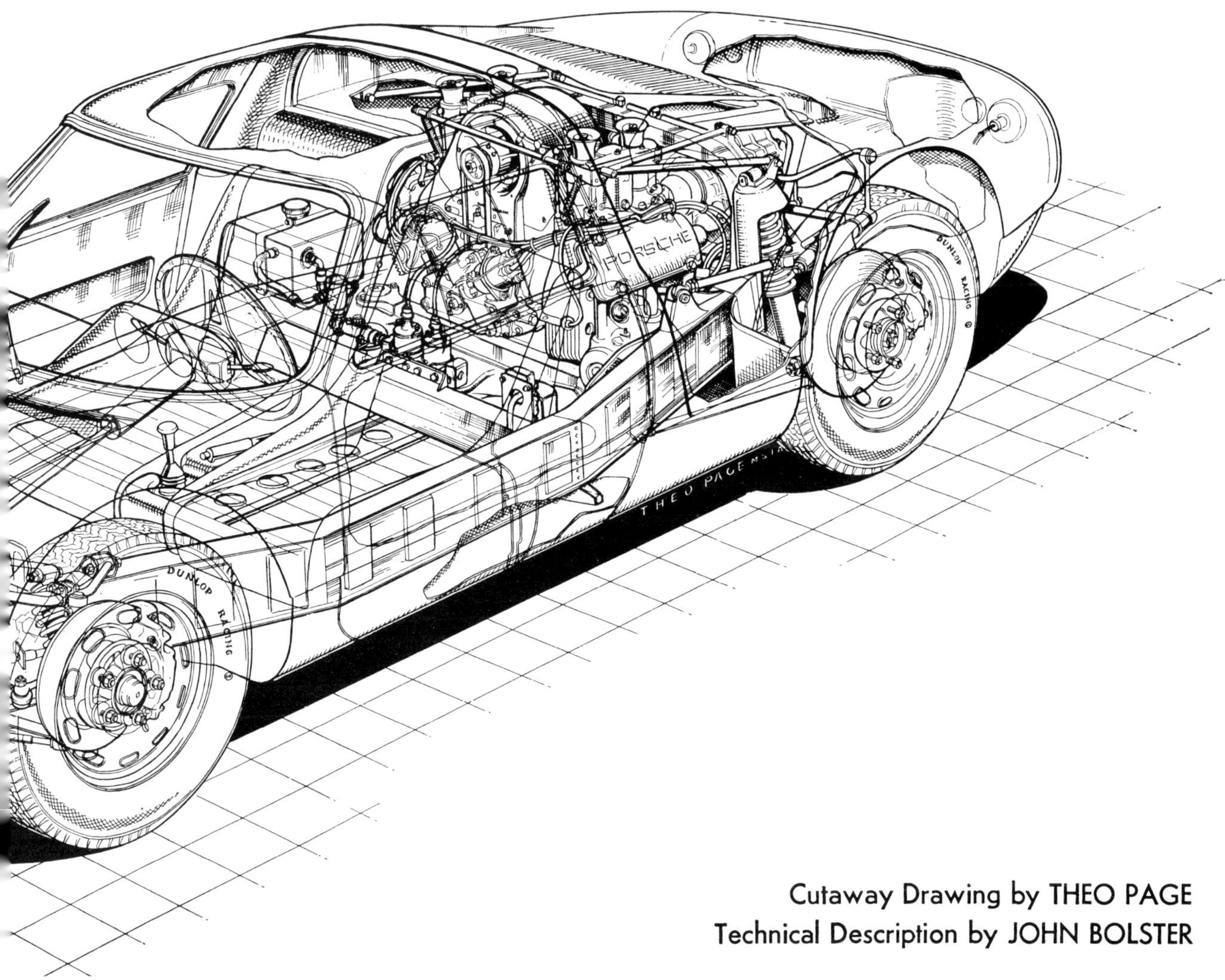

Cutaway Drawing by THEO PAGE
Technical Description by JOHN BOLSTER

New to Porsche: Engine amidships in a closed glass-fibre body. The 904 GTS is intended for the serious amateur racing driver with Le Mans or the Targa Florio as possible fields of endeavour

Porsche 904 Carrera GTS

HOT ROD FROM ZUFFENHAUSEN

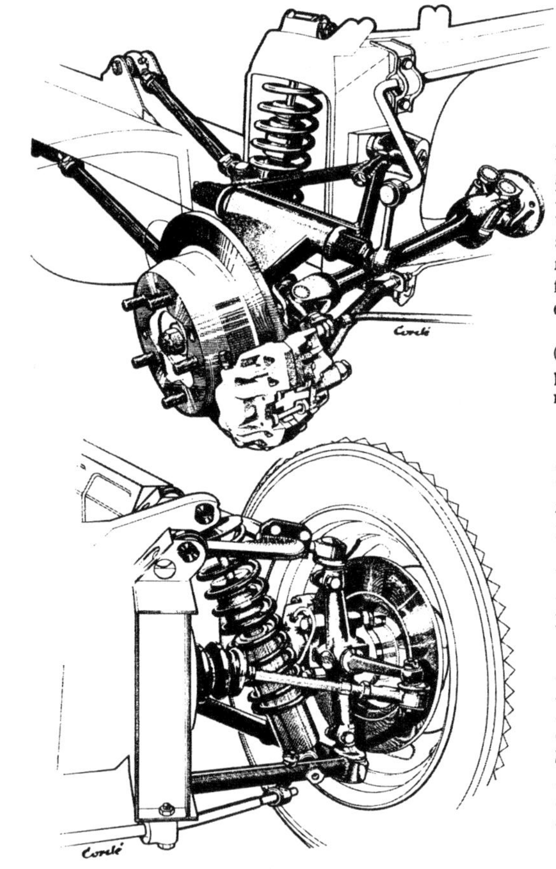

DEPARTURE from a number of " house rules " is evident in the design of the Type 904 GTS Porsche announced last week in Stuttgart and briefly described in *Autocar*. For the first time Porsche are offering to their paying customers a car with a separate chassis frame and a glass-reinforced plastic body. It is also the first closed Porsche to be sold with the engine situated between the passengers and the final drive. Priced at 29,700Dm (£2,670) at the factory, it is hoped to produce a dozen cars for the American market before Christmas and 100 vehicles by the end of next April. The 1,966 c.c. flat four engine will be available in two stages of tune, the standard version giving 155 b.h.p. net at 6,400 r.p.m. and the sports racing version producing 180 b.h.p. at 7,000 r.p.m.

The relationship of this new car to the Porsche range seems much the same as that of the Ferrari GTO to its companion cars, in that it is a sports racing coupé which private owners can enter for such epics as the Targa Florio or Le Mans with some chance of success. There is room in the frame for either the six-cylinder or eight-cylinder units if the factory should choose to race these in prototype cars at a later date.

A good deal of development has been necessary to increase the output of the 587/3 engine for the Porsche 904. The size of the oil radiator on Porsche racing cars indicates that the lubricating oil is used as a cooling medium to conduct heat from the cylinder walls, which gets there by way of the piston crown and piston rings. To further improve heat dissipation, the depth of the cylinder finning of the Type 587 unit has been increased from 7mm to 20mm.

Following previous Carrera practice, a twin-choke Weber carburettor feeds each pair of cylinders, the choke diameter in this case being 46mm. The

Left above: Bearing a close resemblance to that of the eight-cylinder RS Spyder, the rear suspension has equal length transverse links and unequal length trailing arms to impart rear-wheel steering effect. Left: The unequal length wishbone front suspension is quite conventional and follows the pattern set by Porsche for the Grand Prix car. This is the first racing Porsche with plain disc brakes

Below left: Interior finish of the demonstration cars was stark in the extreme. The instrument panel and facia are a one-piece moulding with a padded edge. Below right: A large, light-gauge pressing supports the exhaust system and the rear of the body. In an accident it would also help to protect the gearbox

Porsche 904 Carrera GTS . . .

inlet and exhaust valves are 49mm and 43mm in diameter respectively. Rather extreme valve timing is used on the racing engine, the inlet valves opening 97deg. before t.d.c. and closing 93deg. after b.d.c.; the exhaust valves open 88deg. before b.d.c. and close 60deg. after t.d.c. The flattest part of the torque curve has a narrow range between 4,000 r.p.m. and 6,000 r.p.m.

Basis of the 904 chassis frame is two deep box-form members fabricated from sheet, tied together with hollow, box-form bulkheads at either end and a pair of transverse boxed cross-members amidships. The side members are bowed outwards to enable the passengers to sit low down between them, and to provide leg clearance the forward transverse member is underslung. Although rigid in itself, the chassis is further stiffened by the body, which is bolted and permanently bonded, with adhesive, to the side members.

As mentioned briefly last week the suspension is the same as that of the Grand Prix cars, bearing a close resemblance to the eight-cylinder Spyder which made a first appearance in the Nürburgring 1000km race. At the front it is a straightforward, unequal length wishbone layout with tubular links, the wheelpost swivelling on ball joints. The axis of the chassis pivot points is "toed-in" relative to the centreline of the car to prevent nose-dive when braking. The rear wheels swing on equal length triangulated links with the apex inboard; parallel trailing arms provide fore and aft location of the wheels, the upper arm being shorter to provide a small amount of rear wheel steering. Co-axial coil spring and damper units are used and there are anti-roll bars front and rear.

The drive is taken through a dry plate clutch to the two-shaft, all independent five-speed gear box with Porsche synchromesh on all forward speeds. The solid drive shafts have joggled Hooke joints at the inboard end to take up variations in length due to the suspension geometry. This joggled joint also eliminates the need for splines which tend to lock-up under load.

Porsche-ATE outboard disc brakes with separate front and rear circuits are used. This design, manufactured in Germany under licence from Girling, has a small drum brake incorporated in the rear discs to provide an efficient parking brake.

Heinkel Flugzeugbau, who build aeroplanes, not three-wheeler cars, are responsible for moulding the body. It is laid up in moulds using Gevetex glass fibre and Ranatal resin, produced by Badischer Anilin und Soda Fabrik. Basically it is in three units, the cab and undertray, which are first moulded and then bolted and glued to the chassis frame, and the detachable nose and tail units.

Two-piece Perspex side windows are fitted, the rear portion sliding. Air intake grilles form a pressure area and the rear quarter panels pick up cooling air for the rear brakes. The headlamps are set well back in the long sloping nose, necessitating long clear-plastic fairings.

A demonstration of three 904s was arranged on the Solitude circuit, the two works drivers, Edgar Barth and Herbert Linge giving rides round the circuit, which was open to normal traffic, although there was police control at all junction. Unfortunately it was impossible to drive the cars, which was probably as well in view of the prevailing wet weather, but a fast lap sitting beside Barth was a very convincing demonstration of the stability of the car on slippery surfaces. On the long winding straight at the back of the circuit 200 k.p.h. (124 m.p.h.) came up quite readily in fourth gear and at this speed the car was completely steady. Accelerating hard in second and third through fast bends at speeds in the 60 to 90 m.p.h. range gave a similar feeling of security and one's initial feeling of trepidation quickly changed to one of enjoyment.

Porsche say that the car is not intended for airfield racing, but can compete over the roughest and toughest going, such as one finds in the Targa Florio. To this end, 15in. wheels are fitted, but the ground clearance with standard tyres seems inadequate and one can visualize the undertray suffering on some circuits.

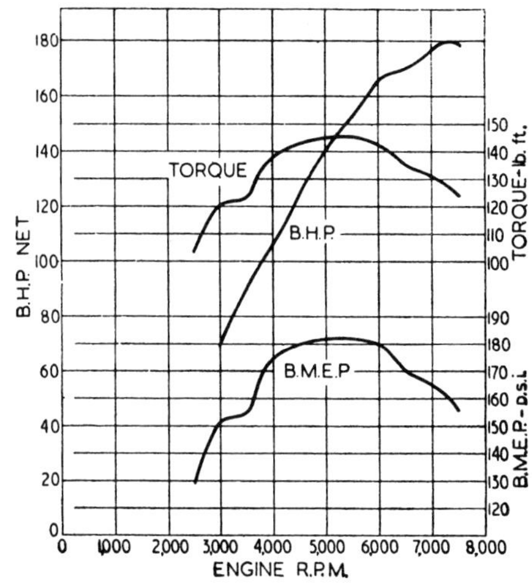

Power and acceleration graphs supplied by the manufacturer indicate that despite the peaky power curve, the five-speed gearbox gives impressive acceleration up to a maximum of 162 m.p.h. in about one minute

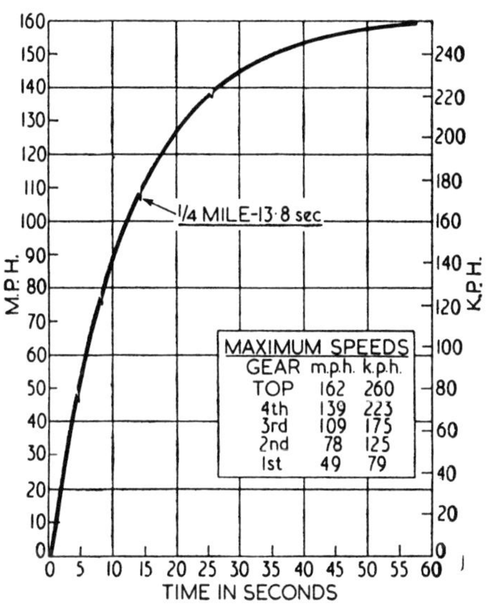

Specification

ENGINE (rear-mounted, air-cooled)

No. of cylinders	Four, horizontally opposed
Bore	92mm (3·62in.)
Stroke	74mm (2·91in.)
Displacement	1,966 c.c. (120 cu in.)
Valve operation	Twin overhead camshafts
Compression ratio	9·8 to 1

Racing Engine:

Max b.h.p. (net)	180 at 7,000 r.p.m.
Max. torque (net)	145 lb ft at 5,000 r.p.m.
Max. b.m.e.p. (net)	182 p.s.i. at 5,000 r.p.m.

Standard Engine:

Max b.h.p. (net)	155 at 6,400 r.p.m.
Max torque (net)	137 lb. ft. at 4,800 r.p.m.
Max b.m.e.p. (net)	172 p.s.i. at 4,800 r.p.m.
Carburettors	2 Weber dual-choke down-draught
Tank capacity	24·1 Imp. gallons (110 litres)
Cooling system	Air-cooled by centrifugal fan

TRANSMISSION

Clutch	Single dry plate
Gearbox	Five-speed, all-synchromesh
Overall gear ratios (racing)	Top 3·36; 4th 4·26; 3rd 5·39; 2nd 7·46; 1st 11·7
(standard)	Top 3·00; 4th 3·94; 3rd 5·39; 2nd 7·46; 1st 11·7
Final drive	Spiral bevel, 4·43 to 1

CHASSIS

Frame	Box-section ladder frame with glass fibre reinforced plastic body
Brakes	Porsche-ATE dual circuit disc front and rear
Suspension: front	Independent, double wishbones, co-axial coil springs and dampers, anti-roll bar
Suspension: rear	Independent, double transverse links and radius arms, co-axial coil springs and dampers, anti-roll bar
Wheels	Pressed steel disc, 15in. dia.
Tyre size	165—15in. (standard) 5·50—15in. front, 6·00—15in. rear (racing)
Steering	Rack and pinion

DIMENSIONS (Manufacturer's Figures)

Wheelbase	7ft 6·75in. (230cm)
Track: front	4ft 3·6in. (131cm)
Track: rear	4ft 3·6in. (131cm)
Overall length	13ft 6in. (409cm)
Overall width	5ft 0·75in. (154cm)
Overall height (unladen)	3ft 5·5in. (106cm)
Ground clearance (laden)	4·7in. (12cm)

PORSCHE TYPE 904 GT

Porsche isn't officially in road racing these days, but a lot of interesting things are coming out of Stuttgart!

WHILE GIVING NO INDICATION of a return to road racing, Porsche's renewed interest in the sport seems to be gathering momentum like a rolling stone. A two-liter production version has barely been announced when up pops the beautiful GT coupe pictured on these pages. It's what might be described as a 1970 Carrera, combining a 180-horsepower, two-liter, *four*-cylinder in RS configuration (engine forward of the transaxle) with full-independent— not swing-axle — suspension, coil/shocks, and disc brakes all around. The transmisison is a five-speed, close-ratio, full-synchro unit. Porsche explains this latest brainchild as acknowledging the wants of the competition-minded private owner, offering him a car for the Targa Florio, high-speed rallies, or even sprint racing. There seems little doubt that it will accomplish the job, as the technical specifications indicate. The tab, in case you have your check book out is 29,700 DM or roughly $9000, F.O.B. Stuttgart; not at all unrealistic for a top-flight GT.

Basis for the two-liter car is a tall, box-section pair of rails to which full independent suspension is fixed. Steering is rack-and-pinion, brakes are outboard-mounted disc-type. Features of the sleek body include an inset rear window and doors that cut down into the rockers and up into the roof. Many points of the design show consideration toward production in fair quantities.

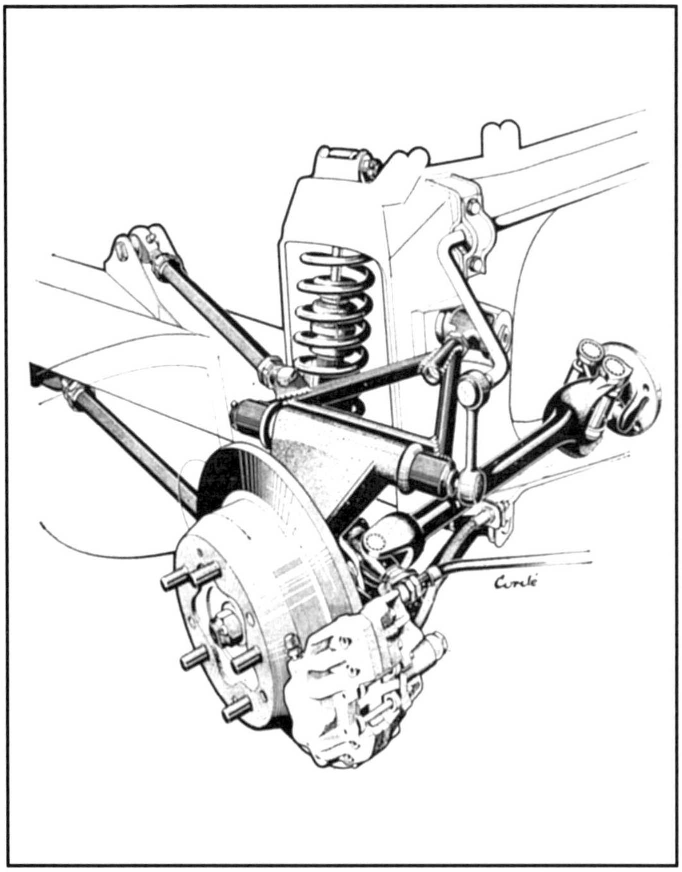

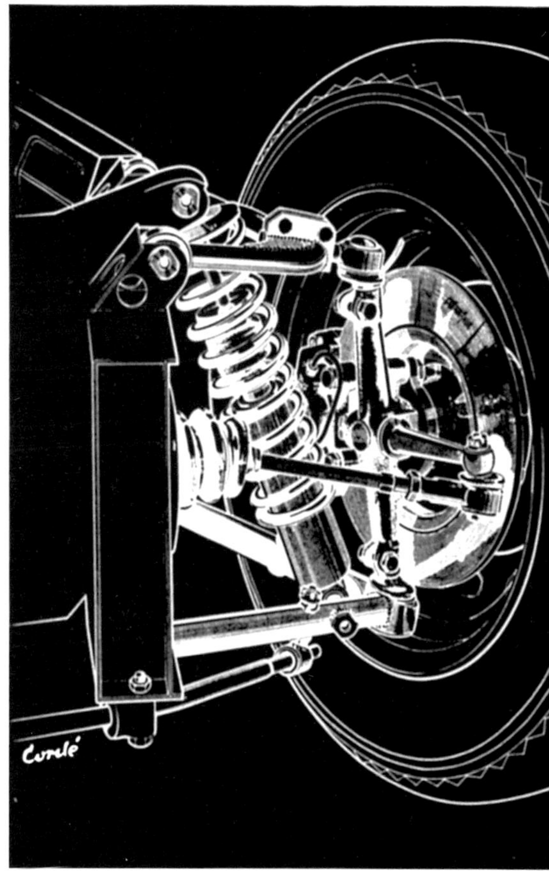

by BERNARD CAHIER

PORSCHE have been accused recently of becoming too conservative. It is true that until the presentation at Frankfurt of their new 2-litre six-cylinder 901 their line had remained basically unchanged for the past 10 years. Fortunately, in November last year they came up with their new Type 904 GTS, a car showing strikingly that this famous though small firm had lost neither their touch nor their imagination. Chassis, body and suspension are all new.

Rather than fit the elaborate and expensive eight-cylinder engine (used for both formula 1 and sports car racing) in this new car, the engineers decided to stick to their faithful four-cylinder, four-o.h.c., Carrera unit; a new design for the head, different-type pistons and a new camshaft have raised the 160 b.h.p. of the racing GT version of this engine to 180 h.p. Din (200 b.h.p. S.A.E.) as well as improving the maximum torque from 125 lb. ft. at 4,800 r.p.m. to 148 at 5,000 r.p.m.

Around the engine has been built a lighter car with more up-to-date suspension and a new streamlined body capable of giving those precious extra miles per hour in top speed. Instead of making an expensive tubular-type chassis for the 904, Porsche made a simple but very sturdy box-section frame. The typical Porsche suspension, torsion bars, trailing arms and swing axle at the rear, was abandoned in favour of all-round independent suspension directly derived from their Grand Prix cars.

To save weight and money, a glass-fibre body was selected for the first time in Porsche history. Designed by the eldest son of Ferry Porsche, this is composed of two main sections. The first, in one moulded unit, forms the front end of the car, the cockpit and the top; the second unit includes the whole rear end, roof and wings. The front section is fixed while the rear is quickly removable, giving full access to the engine and rear suspension.

Knowing the importance which Porsche give to streamlining, one can be sure that past racing experience has been used to good advantage, as well as careful wind tunnel studies. On the scales the 904 shows a weight of 1,430 lb with 11 quarts of oil in the tank (it is dry-sump). This is about 250 lb fewer than their 1963 G.T. competition car, and 600 fewer than the production Carrera GS. This

ON TEST

PORSCHE 904 GTS

Superbly clean-lined, even to the recessed fuel filler cap and door catches, the Porsche wastes none of its 180 b.h.p. The Bosch bus-type single-blade parallel-arm windscreen wiper is seen clearly in the head-on view. The blade remains vertical throughout its sweep

The view through the vertical, shrouded rear window can scarcely be described as panoramic

The two-piece glass-fibre body has the fore part rigidly mounted to the box-section chassis frame and the rear part hinged for access to the engine

what was most important was to see that this car could be driven at very high speeds on wet roads in a perfectly relaxed manner. This was owed, of course, to the good general handling qualities of the car, to its fine steering and well-balanced, powerful disc brakes; but it was also owed, during this wet test, to the best windscreen wiper (a single blade) I have ever seen on any car. It is a Bosch parallel-arm type as used on buses. Windscreen wipers are becoming terribly inefficient at speeds over 70 or 80 m.p.h.

The shape of the 904 (it is very low with a total height of 41 in.) is such that most of the water thrown back at it by other cars does not seem to hit the windscreen, but goes above it.

When I took over the 904 GTS in the Porsche competitions department, it was the first time I had actually seen the car. One disappointment was that the finish of the inside is not up to the standards one expects of this *marque*. The car I drove was one of the very first prototypes; but

Left: Rear suspension that was once familiar on the formula I G.P. cars, and the four-o.h.c. flat-four 2-litre Carrera engine are very accessible when the rear section of the body is removed. Right: Functional and unadorned, the cockpit is clearly that of a competition car. Seats are not adjustable fore-and-aft, the driving position being tailored by adjusting the pedals and steering wheel for reach

weight saving, the extra horsepower and the extremely good shape of the car make it easy to see why the new 904 is so much faster than the not-so-slow 1963 GT model.

This assumption was quickly confirmed for me when Porsche racing manager, von Hanstein, arranged a test of this exciting newcomer for *Autocar*. This test was made around Stuttgart, on the open road, the Solitude circuit and, of course, on the *autobahn*. During my two days' stay the weather was not especially pleasant, with rain and snow too often interfering, but it proved to me that the 904 is a fantastic car in the wet; that you can get quite wet in it; and that an unwanted inside breeze can keep your feet very cold in wintertime.

Mind you, there was a heater, a luxury for a competition car, but it seemed to work mostly around the upper part of my body. But those are minor comfort details and production models, although much better, are still not as well finished as they should be. Granted it is a competition car, but some of them are being sold for road use, with a milder engine and quieter exhaust.

Getting into a 41in. high coupé imposes some problems for girls wearing tight skirts, but a reasonably agile man will not find the operation too painful. After a little effort one is rewarded, as the sitting-cum-driving position is very good, and the cockpit roomy enough to accommodate six-footers comfortably. The seats are not adjustable, as they are part of the moulded front unit of the car's body. Therefore, a suitable driving position is achieved by adjusting the distance of the three pedals via a short handbrake-type lever. The steering column is also adjustable for reach.

Concluded on page 113 ▶

PHOTOGRAPH: JULIUS WEITMANN

Road Research Report:

PORSCHE 904

A two-liter dual-purpose Porsche with the handling of a Formula One car and the performance of a GTO.

Our formal introduction to the Porsche 904 should have been a disappointment. We were given the third 904 ever built, an ancient factory prototype that had been a mechanical guinea pig for the engineering department. It was spottily painted a dull, military shade of khaki, with virgin (if sullied) fiberglass showing through in places. It looked like some kind of crazy, knee-high reconnaissance vehicle that had failed its tests at the Aberdeen proving grounds. What obviously should have been a museum piece was to be our test car.

Baron Huschke von Hanstein, Porsche's racing-, press-, and public relations-manager, was apologetic. Despite the fact that 904s were being turned out at the rate of almost one-a-day, eager customers were snapping them up at a pace production facilities couldn't match. We grandly waved aside his reservations; after all, if a lowly farm tractor had been good enough for von Hanstein's decampment from East Germany many years ago. Any 904 was better than no 904 at all.

Starting our check-out procedure, the good Baron popped easily into the low slung coupe and beckoned us to follow. He's a very lithe fellow for his 53 years.

The doors run the full height of the car—even curving into the roof—but opening them reveals a foot-high chassis side rail that blocks off about a third of the aperture. With practice, getting in and out can be done quickly but never gracefully.

By the time we were aboard, von Hanstein was ready to fire it up. The fuel pump was clicking, the dual ignition switches were pushed home and a touch of the key-operated starter was all that was needed to fill the tiny cockpit with noise. Even with street mufflers and air cleaners, the 904 comes on like the loudest part of a war movie sound track. That old Porsche trait, spitting back through the carburetors, is still there, as was a certain amount of lusty backfiring. We could swear we heard every bearing, shaft, gearset, tappet, and reciprocating part in the engine. Underway, these sounds meld into a furry-edged buzz—then a hollow, throaty growl—and finally reach a richly harmonic, high-pitched drone as the 7000 rpm red-line is approached. The loudest noise at 60 mph is your heart pounding in anticipation, and normal conversation is utterly impossible at 100 mph.

As we banged, popped and stuttered toward the factory's heavily guarded exit, we had a chance to mentally review our notes on the car.

Technically, there is little novelty in the 904. Except for the chassis and body it is basically an amalgam of everything the factory has learned from racing Spyders, Carreras and single seaters.

The chassis, which merely ties all the bits together, is a simple rail job. No space frames, no unit-body or monocoque construction—just a pair of deep, more-or-less box section rails that might be straight out of Frank Kurtis' design book. There are a few stamped indentations and cross-braces, but to the Sunday-supplement engineer's eye, it looks remarkably limber.

The body was designed by chassis engineer Schroeder and Ferdinand Porsche, Jr. (grandson of the late Dr. Ferdinand). It appeals to one's baser instincts rather than to one's aesthetic sense. There have been dark rumors that its aerodynamics and top speed are improved by leaving the detachable rear body section at home, and as for originality, it shows the influence of Carlo Abarth's 1958 designs for Porsche.

The factory claims the body—which is bonded directly to the frame rails—lends structural strength to

PORSCHE 904

the chassis. It is made of an ersatz fiberglass (called Pantal) by the BASF firm in Ludwigshâfen and molded by the Heinkel works. It feels rough and heavy by comparison to its more sophisticated American and English counterparts and seems to shred easily in minor racing shunts. Difficulties in controlling wall thickness are blamed for the 904s' excessive weight at Sebring where the heaviest example weighed 1710 lbs; the factory now promises a delivered weight of under 1540 lbs. As far as the FIA is concerned, it could weigh as little as 1360 lbs before it would be under its homologated weight less the allowable 5%. Alternate bodies are permissible . . . anyone care to try an aluminum-bodied roadster?

The exterior finish (silver—other colors on request) is good and our argument about chassis flexibility is shaken to pieces by one empirical spin around the block—it rides as solidly as a tree stump.

The engine is the tried-and-true Porsche two-liter, air-cooled flat-four with four overhead cams. (Incidentally, they've dropped all that mystique about chrome-plated bores and roller bearings in favor of sprayed-steel liners in aluminum cylinders and plain bearings for con-rods and mains, like the Super 90 and 356SC.) For the first time since the original series of 50 Porsches were built in Austria, the engine is ahead of the axle in a Porsche production car. Basically the same powerplant as the late-model Spyder and Carrera 2 engines, it is rated at 180 bhp at 7200 rpm with 144.5 lbs/ft of torque at 5000 rpm by the conservative DIN system. This engine differs from previous Porsche two-liters in having wilder cams, bigger valves and more fin area. Detuned for street use (with air cleaners and muffler) it produces 155 DIN bhp at 6900 rpm and 124 lbs/ft of torque, again at 5000 rpm.

There just happens to be room in the 904's engine bay for anything up to an aluminum 377 Chevy V-8, though experimentation thus far has been confined to a two-liter version of the Formula One car's flat-eight. Word around the back room is that the flat-eight is falling from favor, and that eventually the 901's flat-six—tuned to over 190 bhp—will be available in the 904.

The suspension is nearly identical in geometry and dynamics to the last (1962) Porsche Grand Prix cars, though some changes have been dictated by the sports car's different characteristics, and for easier "mass" production. It consists of a conventional double wishbone system front suspension and a complicated four-link rear suspension—both with telescopic shocks and coil springs. There's not a torsion bar in sight, save for the front and rear anti-rollbars. All the struts are

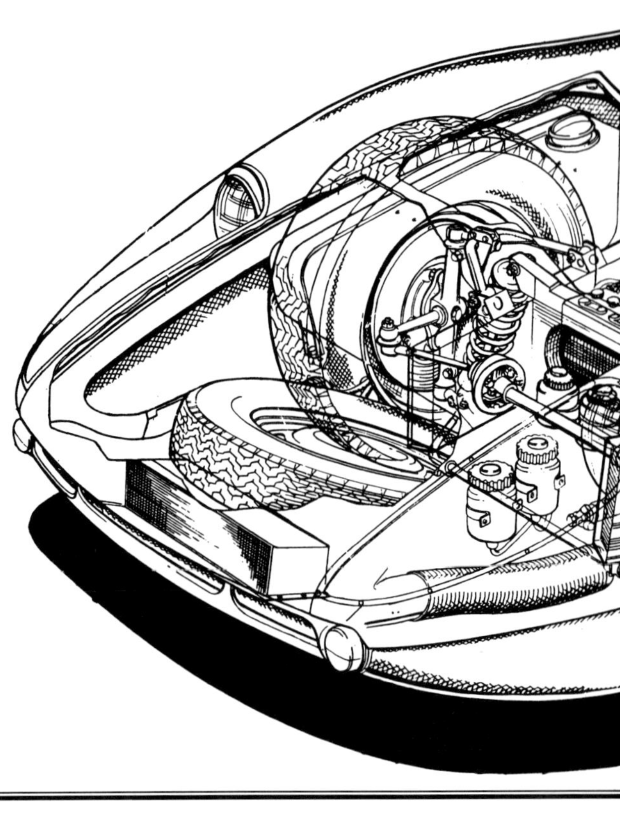

Simplicity and craftsmanship are Porsche traits

tubular steel fabrications, while the front uprights are forged and the rear hub carriers are cast in light alloy. The rear suspension employs two long trailing arms, a reversed upper wishbone and a slightly-trailing lower link. Light alloy trailing arms are optional.

Single universal joints are used as the outboard ends of the open halfshafts; the inboard ends have two sets of U-joints that give a "double knee action" to accommodate small changes in track. This ingenious and positive system does what sliding splines and Metalastik joints do for other cars (though it would not be suitable for anything as powerful as a Cobra, for example). Unlike the 356C Porsches, the 904's suspension system is lubricated-for-life except for the leading ends (Heim joints) of the rear trailing arms.

The steering is a new-to-Porsche rack-and-pinion design from ZF (prefiguring the 901's steering) while the brakes are Dunlop-Ate four-wheel discs, mounted outboard. Like the 356C, the rear brakes have a small drum within the disc hub, to be used as an emergency or parking brake.

The gearbox is really the newest—and one of the best—features of the car (and, again, prefigures the 901's almost-identical transmission). It is a five-speed all-synchromesh unit, using the justly-famous Porsche synchro system, in a new casing with a host of improvements over the Spyders' five-speed. (The old five-speed, never used in a production car, was basically the 356's four-speed with first in place of reverse and reverse hung off the nose piece in an auxiliary case.) A ZF cam-type limited slip differential is standard equipment on the 904, as is a 4.428-to-one ring and pinion. Sixteen gearsets are available with ratios ranging from 2.83:1 to .760:1 (six to 22 mph per thousand rpm with the standard 165-15 Dunlop SP tires). The gearbox selector pattern is artfully arranged with a normal four-speed H-pattern at center right; starting and reverse gears are to the left, beyond a spring-loaded detent. Reverse (forward) has an additional detent (down), so it takes the novice some time to get moving astern.

The dashboard (in black wrinkle-finish fiberglass) is unusually complete with instruments and controls. Centered in the nacelle is a 9000 rpm tach flanked by a 280 kmh (190 mph) speedometer on the right and a combination oil pressure/oil temperature/fuel gauge on the left. Various warning lights tell the driver if the heater is off, the generator is *kaputt,* the high beams are on, the parking lights are on, or if the turn-indicator blinkers are on. Two stalks on either side of the steering column control the horn, headlight dimmers, directional lights and auxiliary driving lights. There's a push-button windshield washer and an absolutely

Text continued on page 108; specifications overleaf

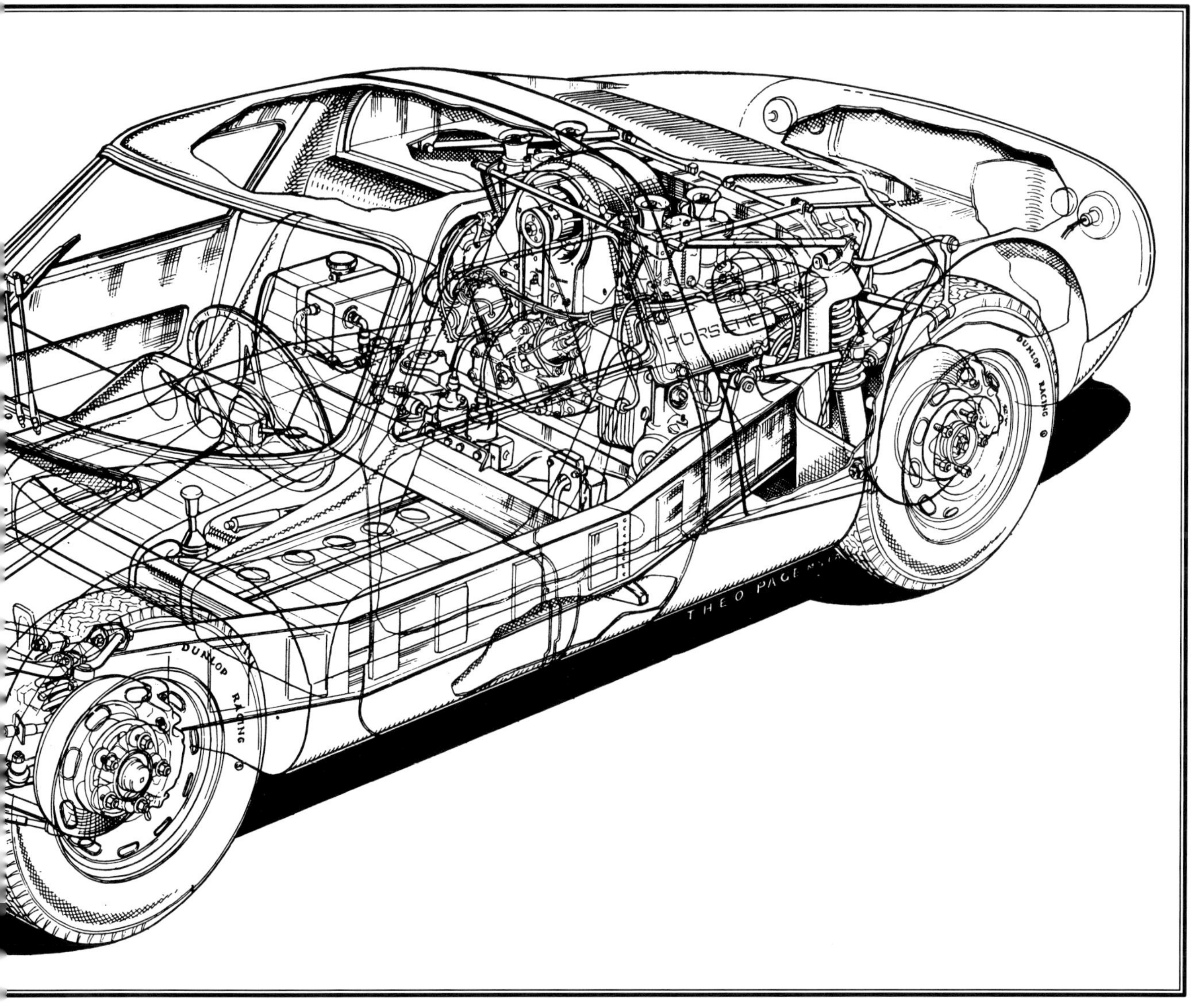

Road Research Report: Porsche 904

Importer: Porsche of America Corp.
107 Wren Avenue
Teaneck, New Jersey

PRICES

Price as tested: $7425 F.O.B. Stuttgart

ENGINE

Air-cooled flat-four, light alloy block, 5 main bearings

Bore x Stroke	3.62 x 2.91 in, 92 x 74 mm
Displacement	120 cu in, 1966 cc
Compression ratio	9.8 to one
Carburetion	Two Weber dual-throat 46 IDM
Valve gear	Four overhead camshafts
Valve diameter	Intake 1.935 in, exhaust 1.69 in
Valve lift	Intake 0.484 in, exhaust 0.508 in
Valve timing:	
Intake opens	79° BTC
Intake closes	93° ABC
Exhaust opens	88° BBC
Exhaust closes	60° ATC
Power (SAE)	198 bhp @ 7200 rpm
Torque	148.3 lbs-ft @ 5000 rpm
Specific power output	1.65 bhp per cu in, 100.4 bhp per liter
Usable range of engine speeds	2000-7500 rpm
Electrical system	12-Volt, 45 amp-hr battery, 300 W generator
Fuel recommended	Super premium
Mileage	15-20 mpg
Range on 29-gallon tank	435-580 miles

DRIVE TRAIN

Clutch	7.9-inch single dry plate
Transmission	5-speed all-synchro

Gear	Ratio	Over-all	mph/1000 rpm	Max mph
Rev	2.610	11.58	—6.65	—50.0
1st	2.643	11.66	6.6	49.5
2nd	1.550	6.87	11.2	84.0
3rd	1.125	4.99	15.4	115.8
4th	0.889	3.94	19.6	147.0
5th	0.759	3.36	22.8	160.0

Final drive ratio .... 4.428 to one

CHASSIS

Double rail frame bonded to fiberglas body.

Wheelbase	90.6 in
Track	F 51.7, R 51.6 in
Length	161 in
Width	60.6 in
Height	42 in
Ground clearance	5.0 in
Dry weight	1435 lbs
Curb weight	1540 lbs
Test weight	1800 lbs
Weight distribution front/rear	42/58 %
Pounds per bhp (test weight)	9.10

Suspension F: Ind., unequal-length wishbones and coil springs, stabilizer bar.
R: Ind., four-link with radius arms (reversed upper wishbone and single lower arm) and coil springs, stabilizer bar.

Brakes....10.8-in discs front, 11.3-in discs rear, 376 sq in swept area

Steering	Rack and pinion
Turns, lock to lock	2.0
Turning circle	44 ft
Tires	Dunlop SP 165 x 15
Revs per mile	820

MAINTENANCE

Crankcase capacity	10.5 qts (dry sump)
Oil change interval	3000 miles
Grease fittings	4

ACCELERATION

Zero to	Seconds
30 mph	2.0
40 mph	2.8
50 mph	4.2
60 mph	5.3
70 mph	6.5
80 mph	8.1
90 mph	10.2
100 mph	12.2
Standing ¼ mile	108.5 mph in 14.2

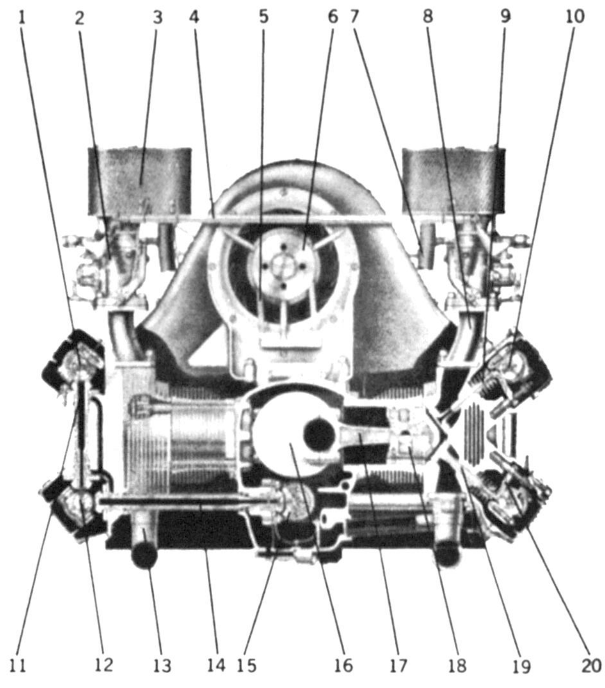

1/10 SCALE

Street engine: 155 DIN bhp, 175 SAE bhp. (1, 10) Intake cams, (2) Solex 44 PII-4 carburetor, (3) Air filter, (4) Throttle rod, (5) Fanbelt, (6) Generator, (7) Carburetor heating tube, (8) Intake manifold, (9) Intake valve, (11) Short jackshaft, (12) Exhaust cam, (13) Exhaust manifold, (14) Long jackshaft, (15) Jackshaft drive gear, (16) Crankshaft, (17) Connecting rod, (18) Piston, (19) Exhaust valve, (20) Cam follower/finger.

STEERING BEHAVIOR

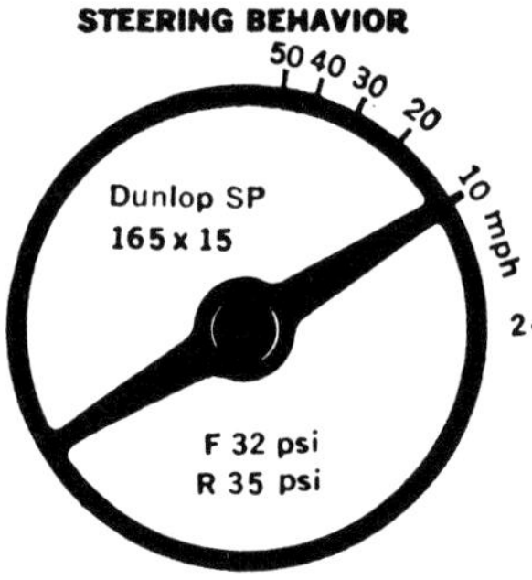

Wheel position to maintain 400-foot circle at speeds indicated.

ENGINE FLEXIBILITY

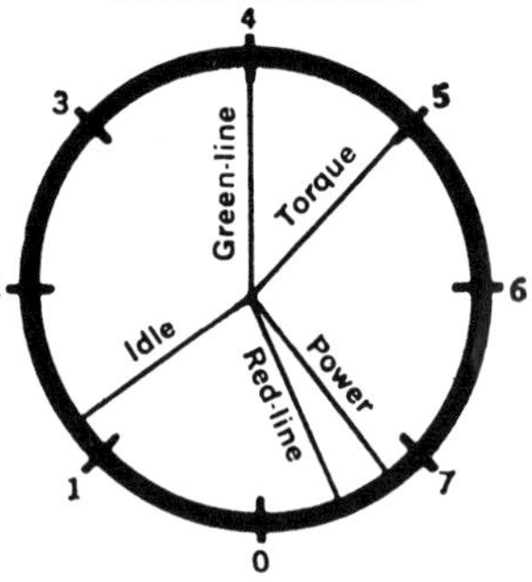

RPM in thousands

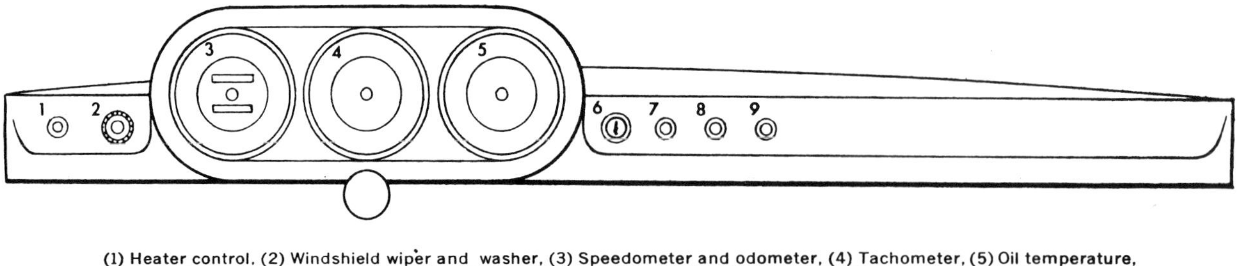

(1) Heater control, (2) Windshield wiper and washer, (3) Speedometer and odometer, (4) Tachometer, (5) Oil temperature, oil pressure and fuel gauge, (6) Ignition key and starter switch, (7) Headlight switch, (8) and (9) Dual ignition switches.

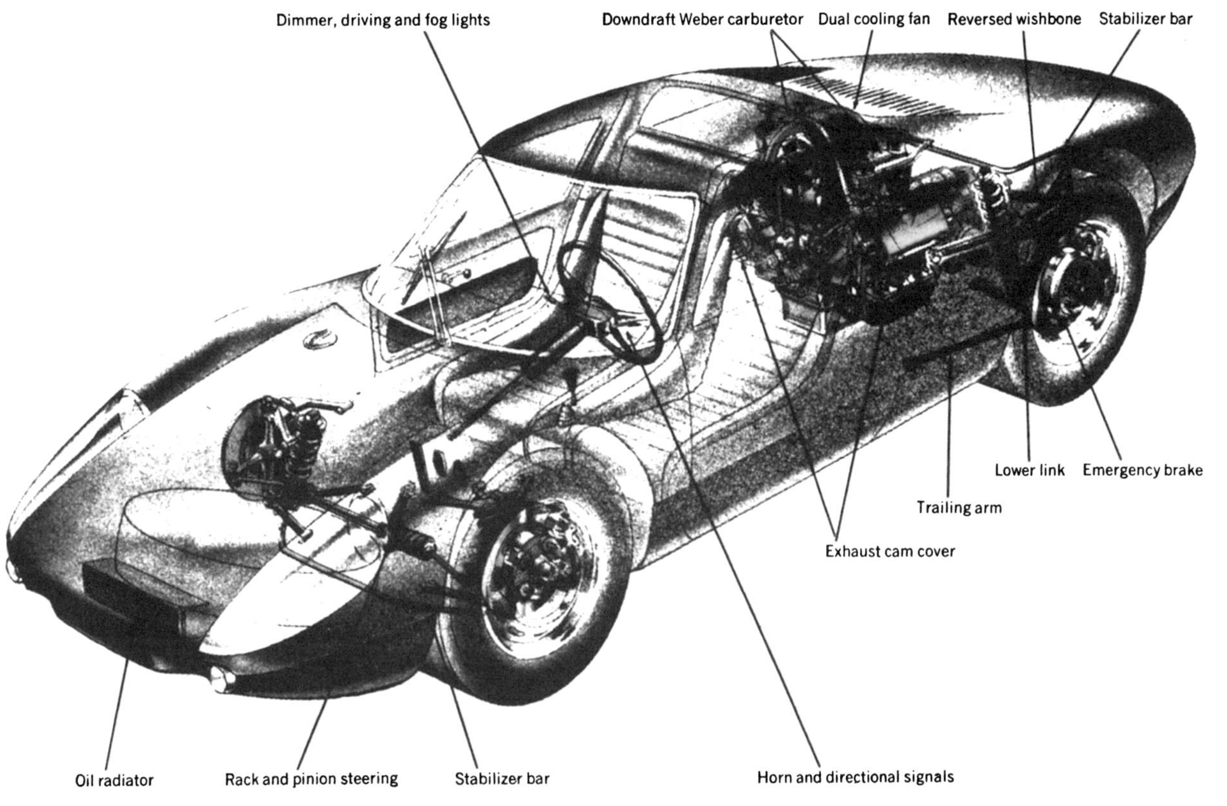
Dimmer, driving and fog lights
Downdraft Weber carburetor
Dual cooling fan
Reversed wishbone
Stabilizer bar
Lower link
Emergency brake
Trailing arm
Exhaust cam cover
Oil radiator
Rack and pinion steering
Stabilizer bar
Horn and directional signals

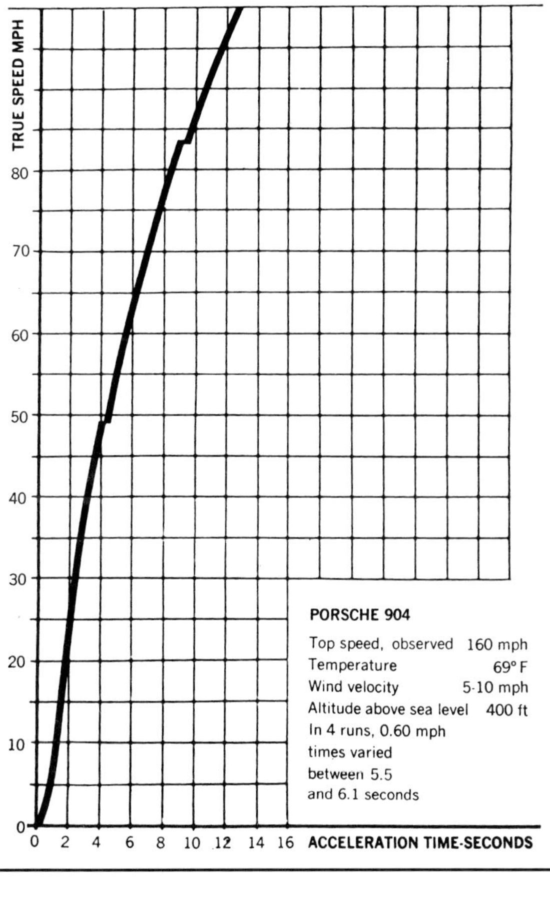
TRUE SPEED MPH
0
10
20
30
40
50
60
70
80
0 2 4 6 8 10 12 14 16
ACCELERATION TIME-SECONDS
PORSCHE 904
Top speed, observed 160 mph
Temperature 69° F
Wind velocity 5-10 mph
Altitude above sea level 400 ft
In 4 runs, 0-60 mph
times varied
between 5.5
and 6.1 seconds

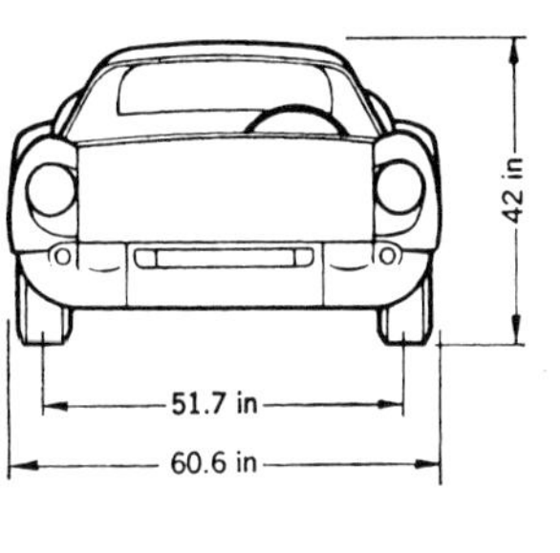
42 in
51.7 in
60.6 in

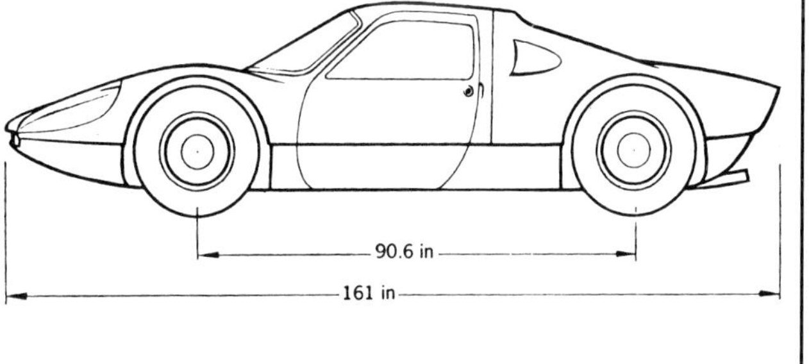
90.6 in
161 in

PORSCHE 904 *CONTINUED*

enormous truck-type single-blade windshield wiper with a "pantograph" strut to keep the blade vertical as it sweeps through its wide swath.

The molded foam rubber seats are stationary while the pedals and steering column are adjustable, like Ford's original (rear-engined) Mustang. With the pedals in their forward-most of the three positions, von Hanstein could neither fully declutch nor achieve maximum braking, as we found out to our horror at the first turn outside the factory gate. This has long since been corrected on the production 904s—we obtained one of the 105 examples that have been built as of this writing, to ascertain this and to better our acceleration figures—but heel-and toeing in the old prototype was difficult for a six-foot-two driver.

A variety of optional-sized seats are available for production 904s to suit drivers of various girths and give enough headroom for up-to six-footers. The seats, covered in blue felt, simply "plug in" to the recesses molded in the back of the cockpit. The interior, as drab as a Populist meeting hall, is surprisingly comfortable (for the driver; the passenger's right foot loses out to a battery box) even liveable for a trip of, say, six hours. There's even a shelf to brace the left foot while whistling around corners.

Ze enchine, I was told, ist chust an oldt Carrerra mohtorr, vis only 130 horrses. Ist much betterr vis toolitah 180 horrsepowerr rracing enchine. Oh sure, only 130 bhp. It didn't seem hardly enough to do any more than pool all our blood along our jellied spine, break our glasses across the bridge of our nose and leave the impression of our belt buckle on our stomach.

Von Hanstein was winding it out so quickly in each of the ultra-close ratio gears that we hardly noticed that we were doing 100 mph only three blocks from the factory. By the time we had taken that in, we were already turning 7000 in fifth and well out of town.

"You're sure President Johnson does this?" we croaked weakly.

"Ja, shoorr—*look!*" He took both hands off the small, wood-rim steering wheel. The 904 tracked along as steadily at 120 mph as a Yankee Clipper. The steering, however, is very sensitive (with only two turns, lock-to-lock): to change lanes he twitched the rim about half-an-inch. As a matter of fact, the 904 can snap from lane to lane at any speed with absolutely no wallowing, yawing or fishtailing . . . and the factory drivers love to demonstrate this to unsuspecting Timid Souls. Lean, needless to say, is imperceptible.

After scaring the living hubris out of us, v. H. gave us the wheel. Fully expecting to stall it dead on the first try, we revved the engine wildly and cautiously eased out the clutch. Just when we thought it wasn't in gear after all, the clutch bit like a steam catapult, the revs dropped to near-zero and we were off in a series of jerks and chirps from the Dunlops. The so-called flywheel is about as thick as the ham in a drug store sandwich, so the revs fall off sharply between shifts—and the tires squeak again. It was only when we got driving the little monster like a full-house competition Corvette—slamming the shift through the gate as fast as our palsied hand could move —that things simmered down and smoothed out. But not even a drag Corvette would run out of revs so quickly in each gear. Although the

904 is not comparable to Pontiac's GTO, it's noteworthy that the Porsche with its long "Le Mans" gearing is only 4/10-second slower to 100 mph than the Pontiac geared to turn 115-mph quarters.

Our wooly exuberance evidently got the better of von Hanstein so we dropped him off at the factory and set out alone for the open reaches of the Solitude circuit, which the GP boys lap at something over 100 mph. It's a lovely and dangerous circuit lined with trees, a couple of hotels and restaurants, ditches, some fencing, poles and Germans. On public roads, its only closed to traffic on race days. The rest of the time, all kinds of native traffic is tearing around in both directions trying to prove something—skill, masculinity, lack of guilt in starting the last war or lack of responsibility in having lost every one since 1870.

At any rate, in the hairpin turns we discovered the expected: the cornering power is fantastic. The tires just lock into the road surface and push the car against seemingly infinite amounts of centrifugal force. In the faster turns we were initially over-cautious. When at last we gave the beast its head, we learned what all those theoretical discussions about a "low polar moment of inertia" added up to—when you lose it, you lose it faster than the reflexes can follow. The factory test drivers later told us you *can't* hang the tail out on a fast bend; you track it around like a GP car or you are too slow . . . or you spin. With almost all the weight of the car within the wheelbase, it breaks loose as quickly as a broom balanced on your palm . . . with the straw end down. Not that it can't be done; the correction must be quick, sure, and metered out in exact amounts at the right time.

The rearward visibility is excellent—much better than one would expect, and better than a pre-1957 Volkswagen, for instance. It suffers only in the left rear quarter where a triangular piece of fiberglass fairing cuts off the view of who's passing you. Not bloody likely. Side and forward vision are as good as, or better than, the late lamented Speedster.

The big question in our mind, and perhaps in those of our readers, was: could this untamed beast find happiness as a pampered street car? (We were ready to start signing over all our travelers' checks . . . as a down payment.) The answer is yes—a highly qualified yes. It *could* be used for boulevard transportation, but the owner/driver would have to be slightly out of his mind.

There are those among us who are going to go out and buy 904s anyway, detuned as little as possible, and tear around the countryside like maniacs. They won't be dissuaded by the awkwardness of getting in and out, by the expense of setting up and maintaining the four-cam engine, by the harsh ride, nor by the insufficient steering lock (several tries are needed to make a U-turn on a narrow street). No sir, they're utter fools, and we wish we were among them.

On the other hand, as a competition car, it's an eminently practical and satisfactory choice. The SCCA recently classified the 904 as a class A Production car on the basis of its potential performance. By that standard it competes with the mighty Cobra. By the FIA's international classification it's an under-two-liter car where its closest rival is the very fast, very fragile Abarth *Due Mille* ("2000"). The 904 is quick and strong enough to acquit itself well in any long distance event, and at $7425 (ex-works) it's a pretty cheap way to get into the Big Time. C/D

PORSCHE 904

IN MID-1963, Ferry Porsche admitted in an interview with *Christophorus* (Porsche's de luxe house magazine) that a new GT car was coming for the 1964 season. Rumors flew thick and fast and in last month's R&T we were able to show the result—the 904—but, as so often happens in early scoops, some specifications came out a bit peculiar and not too accurate. The factory did, finally, have a showing of the Abarth's replacement and it is, as always with Porsche, a highly competitive automobile.

The motive power is the tried and true 2-liter, 4-cyl Carrera engine (now with 180 DIN bhp) but, unlike the Abarth version, has the engine forward of the rear axle. The 5-speed gearbox of the 901 is installed and Ate disc brakes are used all around.

The suspension is mounted on sheet steel outriggers from a basic frame consisting of two oval and rather deep side members. This arrangement is lent stiffness by the lower extremities of the fiberglass body (made by Heinkel) being wrapped around and bonded to the ovals. Besides forming an extremely rigid structure, the 904 chassis-body bonding results in a weight saving of 330 lb which is worth a few horsepower any time.

These improvements, plus better weight distribution, resulted in works driver Linge's allegedly getting the 904 around Nurburg Ring in 9 min 30 sec as compared with his best former time of 9 min 58 sec in the old car. The price is just over $7400 fob Stuttgart and we understand the first 12 should be in the States by the time you read this. Anyone for a little grand touring?

1966 Porsche 906-E

The coming of the Piech Batmobile

by Len Frank

PHOTOGRAPHY BY NEIL NISSING

It doesn't surprise me that, after all this time, I remember that race so well. It was 1966, the last year of the United States Road Racing Championship. The venue, Stardust Raceway, wasn't destined to last much longer. The driver involved, Ken Miles, tragically did not live out the year. Only the memory–and the car–survive.

I don't know why the Porsche Carrera 6 was so unexpected there in the heat and dust of Las Vegas. It had already won its class at Daytona and Sebring earlier that spring, but on the West Coast we were still occupied with visions of homegrown Huffaker Genies and Gurney-Weslake heads, and, of course, the mighty Shelby Cobras.

Those were the days when Lolas and McLarens rumbled behind the Chaparrals, when Fords were beginning to get into greater displacement and political ploys in order to keep up. They were days hot with rumors of racing automatics, mysterious downforces, tires a foot wide, and intake ports big enough to roll a tennis ball through. Mr. Science was coming and there were no limits in sight.

Porsche, the automobile producer, had been around by this time for about 18 years. The company's (excellent) reputation was based in part on actual product and in part on the intimidating aura of The Founder. Dr. Ferdinand Porsche had originally opened a design office and consultancy responsible for such diverse products as the original Auto Union Grand Prix cars, a series of Volkswagen prototypes culminating in the KdF-Wagen project (which became the VW), and the Mercedes T-80 land-speed record contender (which never turned a wheel in anger).

Porsche, the car, was really the product of Ferdinand Anton Ernst Porsche, called Ferry, eldest son of the old Doctor, who, ironically, never rode in a car bearing the family name. The Porsche 356 was based on pre-WW II designs, themselves based on some Teutonic presumptions and misconceptions. Despite that, there were 76,300-odd 356s built over a 16-year period. Starting from the earliest production 356/2 of 1948, these "technically advanced" automobiles featured engine, engine location, methods of construction, and econocar suspension devised for *volks* who had never owned a car.

But, for whatever reason, the little 356 remained a commercial success and that success allowed Porsche to mount a growing racing effort. And there was more success, with wins in every major race. When Porsche began building the Type 550 Spyder variations, the Italian OSCA and Maserati 1500S, the English Lotus and Cooper, never had a chance. Only German Borgwards had a real run, but, despite enormous promise, their impending financial doom left the small-bore classes to the

4-cam Porsches. The first Spyder entered in a race, the 1954 Mille Miglia, won its class.

Your Basic Porsche Spyder used a 4-cylinder big-bore air-cooled *boxermotor* of surpassing complexity—much of it unnecessary. It started with a built-up Hirth crankshaft, with its lovely machinework, roller bearings, and special tool requirement, and proceeded to a series of shafts, shims, and bevel gears to drive the four camshafts. Surely the total number of parts in a Type 547 engine must have made it the most complicated four in modern automotive history. The Fuhrmann-designed *wundermotor* (basis for the Carrera and nearly all Porsche race cars for 10 years) was mid-mounted in the Spyders, bolted to a Schmid-synchro transaxle, the whole held together with a space frame using *lots* of tubes. Colin Chapman would have found it untidy. Suspension was a variation on the Dr. Porsche-decreed theme, double trailing arms front, swing axles rear, the whole sprung with the usual transverse torsion bars. It had been good enough for the original Auto Union at one end of the scale and the humble Volkswagen at the other. Wasn't it good enough for a modern race car? No.

Porsche's late-'50s/early-'60s competition was just beginning to make itself felt. Although the traditional competitors had faded, both Lotus and Cooper had become worrisome from one side, Abarth from the other. While none of the new cars was exactly competitive with the silver Spyders, the threat, clearly, was there. Poopers—Coopers with pushrod Porsches installed—began to kick the Spyders around the tracks in the U.S., and when Ken Miles built a 4-cam-powered Pooper for the West Coast Porsche distributor, it won three quick races against very rough competition, then was retired. Owner John von Neumann insisted that it was *not* retired because it was embarrassing the paying (Spyder) customers. Something *had* to be done.

Porsche kept plugging with the Spyder, a little slowly, a little arrogantly, improving aerodynamics, increasing displacement, experimenting with suspension—nearly a decade's worth of fiddling.

The reason for all of this pre-906 preamble is to reinforce the surprise of seeing the all-new 906. There were those times when, asked for the best car in the world, we, the American enthusiasts, would have answered, "Porsche"; and asked for, say, the best $3000 car, we would have said, "Used Porsche." We assumed that the VW-like dual trailing arm front suspension was the best, that the swing arm rear was *non pareil*, that the mightily complicated 4-cam was right, proper, and necessary, that the high-speed instabilities, unusual cornering angles, and flame-belching exhaust were all elements that *had to be*.

The appearance in 1963 of the Porsche 901 (prototype of the 911) was greeted as heresy by Professors of The Faith. Further heresy a year later was the 904, the first (only?) plastic-bodied Porsche production car. It was the first mid-engined (the 4-cam) production Porsche, the first with modern suspension (unequal-length A-arms front, reversed A-arms rear located by semi-trailing arms, the whole sprung with coils). The 904 was actually a GT built in sufficient quantity to qualify in the U.S. as an A-Production car. When it proved able to run with the dominant Cobras and Corvettes, we forgave it. Fickle?

The prototype 906 went from drawing board to competition in less than two months

Las Vegas was, as usual, too hot. The wind blew too hard, sand powdered across everything, in everything; a poor lubricant indeed. Stardust Raceway now has a freeway on top of it, I think, but in 1966 it was another stop on the old USRRC circuit. The USRRC was the Big Time in those days, with all of the big-and-little-boomers in attendance.

In the first year of the USRRC, George Follmer and Bruce Burness put together a car Porsche *should* have looked at more carefully. Follmer's Lotus 23B was a later interpretation of the Miles/von Neumann Pooper. It surrounded the 2-liter Fuhrmann four with the most advanced sports racing chassis available. It featured frame and suspension technology that allowed Lotus to dominate Formula One, Indianapolis, smaller sports racing and formula classes, big-bore sports racing. Everyone, including, eventually, Porsche, learned lessons from Chapman's lightweights.

Follmer won the first USRRC race in 1965 after Hall's Chaparral dropped out. He won the under 2-liter class in the rest for enough points to give him the overall championship. Wounded outrage from the big-bore owners followed, resulting in a rule change that equated "big" with "important" and assured that no U-2 would ever take the championship again without a string of overall wins. And as the traditional Ferrariesque (read "clumsy") large-displacement racers had given way to Lotus/Cooper types bristling with small-block V-8s, the handling edge that kept the U-2 cars close to the big ones just flat disappeared.

But Follmer was on hand in Las Vegas to defend his championship. The whole cast of characters from the previous year,

except the Chaparrals (Hall had taken his flyers to Europe), was present with some new, important additions. I was there crewing on a Lotus 30, certainly the worst, least successful Lotus racer ever built. We were pitted next to Dan Gurney and his Weslake-Ford/Lola. Down the row was a friend in his 427 Cobra, yonder an obsolete King Cobra, a Lotus 19-Ford, a Porsche Spyder or two, here a Lotus 23 with twin cam Ford, there a McLaren, a Lola, a Genie. In a few short years, the backyard specials, the Birdcage Maseratis, the front-engined Ferraris, the Porsche Spyders had all become passe (along with our Lotus 30): overweight, flabby spearcarriers.

And then Otto Zipper's transporter arrived and a pair of incredibly low, mean, ugly (we thought then) coupes rolled out. Remember the 1957 Plymouth ad slogan: "Suddenly it's 1960." Well, with the unloading of those coupes, suddenly it was 1970. At least.

Dr. Porsche, the elder, had a daughter older by a few years than Dr. Porsche the younger. She married Dr. Anton Piech, who, a few years later, became a founding member of the Porsche design/consultancy. During the postwar period, the Piech side of the family began to grow in corporate importance, displacing some of Dr. Ferdinand Porsche's traditional people in the organization. The eldest son of Louise and Anton, Ferdinand Piech joined Porsche straight from engineering school in 1963. He was responsible for the 6-cylinder version of the 904, did considerable development of the 911 engine, then took over the R&D activities. His first project was a new car, designed to be competitive with Ferrari's first really modern racing car, the 206S Dino. Piech and crew took some shortcuts to success by using Lotus F1 wheels, hubs, suspension uprights, and the more modern 13-in. GP tires that went with them. The car went from drawing board to skidpad in little less than a month. A week after that, it was busy losing to the little Ferrari in a hillclimb. The Ollon-Villars Spyder, the first Piech product, was the starting point for the 906 (as well as for considerable internal struggle at Porsche, which seems to be in progress to this day).

Resentment on the Porsche family side of the camp was certainly to be expected. The 904 was still a winner. Although the 206S Ferrari made a shambles of Porsche's traditional domination of the Hillclimb Championship, 1964 had seen the 904 a winner in the Targa Florio, with class wins at Sebring, Spa, Nurburgring 1000, LeMans, Rheims, Watkins Glen, and a dozen rallies. In 1965, the 904 and/or variations took the GT class at the Nurburgring, a class win at the Daytona Continental, Monza 1000 km, Spa, LeMans, and Bridgehampton, plus more than their share of rallies, hillclimbs, and lesser races.

"Butzi" Porsche, son of Ferry, had done the 904 body design. The car observed most of the spirit of the rules, perhaps the last successful sports-racer (with the possible exception of the Ferrari 365 GTB/4 Daytona) to do so. The 904 was built in Spyder form with Porsche's flat eight and eventually the racing version of the 911 six. There was little doubt that there was life left; the car was only two years old, winning races, with customers waiting—always important at Porsche. But for all this, it still was being relegated to the museum. Butzi Porsche eventually left Porsche, partially as a result of this presumptive insult. Other Porsche regulars moved aside—or out—and the Piech star rose toward its zenith.

Porsche built the car as a tool to do a job, not as a shrine; that status came later

The 904 was originally to have used the 911 six and variations, but the engine wasn't ready in time. By the arrival of the 906, the new engine *was* ready. Differences from the now-familiar 911 engine were all in detail—and all important. Engine cases, and some other castings were magnesium rather than aluminum. Titanium con rods replaced steel, dual ignition replaced single, Weber 46 IDA 3C carbs were substituted for the floatless production Solexes. Chrome-plated cylinder bores were used in the magnesium cylinders. The result was 210 DIN hp from a 2.0-liter engine—about 70 more than the production 911S (which itself had well over the then-magic 1 hp/cu in.). The production engine and race engine had crankshaft design, cam drive, and dry sump oiling in common.

Standing there in the pits at Stardust, waiting for the asphalt to soften and the wind to die, the sound of the 906 ripped above the boom of the V-8s. Scooter Patrick was teamed with Miles. Patrick, like Miles (and Follmer), was a product of West Coast Cal Club/SCCA racing—a hard, aggressive driver/mechanic who went about as fast as anyone without breaking cars. Not too unexpectedly, Jerry Grant in the AAR Lola-Weslake took the pole, our Lotus coughed its pistons up through the Tecalemit-Jackson injection trumpets, and both 906s qualified well. With a V-8 properly on the pole, no one paid particular attention. They were, after all, small-bore cars with engines less than half the size of the average V-8 running.

Ferdinand Piech had learned some important lessons from his hillclimb car, so when laying out the 906—officially called Carrera 6—it was no surprise that the car had a spaceframe rather than the unit/platform of the production Porsches, or the deep, fabricated frame rails of the 904.

Spherical joints replaced rubber bushings once and for all. Wheel width was pushed to the point where advanced tire technology could be utilized. Brakes were

lifted from the late 904s, conventional discs but with dual master cylinders and a balance bar. Wheelbase was the same as the 904 (90.6 in.), but track was considerably wider. Overall width was pushed up over the much wider wheels without aerodynamically inefficient flares. Chassis tubes were used to route oil from engine to cooler or reservoir, depending on temperature. Shades of Lotus and Cooper. Body panels did not add to the chassis stiffness (unlike the 904). Thanks to its greater width and frontal area, the 906 actually was a worse aero package than Butzi Porsche's intuitively designed 904. However, the 906 did have additional front downforce compared to its predecessor.

The Retrospect car, a 906E (one of nine), has a longer nose that replaced the add-on "whisker" spoilers of the original 906. There were also long-tailed cars with a better Cd (0.326 *vs.* 0.335) used at Le Mans. The "E" is for *Einspritz*—Bosch mechanical fuel injection and transistorized ignition that went some way toward civilizing the 906. Horsepower rose to 220 DIN/ 235 SAE at 8100/8200 rpm. Another 100 rpm was available when the 206S Ferrari was in the neighborhood—but not for too long, *bitte*. The factory claims a 177-mph top speed. This is conceivable, but only with plenty of room to run.

Warren Eads' 906E is gorgeous—maybe a little too much so. He's lucky to have it. After its checkered past, *we're* lucky to have it. It could easily have become a vaguely automotive heap at the side of a track somewhere. Briefly, it started its career by running at Daytona with Mitter and Rindt driving. That was 1967. It was the first under 2-liter car at Watkins Glen and Sebring in 1969, and at the Daytona 24-hour in 1970. Fred Baker, a driver of some note, campaigned it against Joe Buzzetta in a similar car. What one didn't win, the other did. Dick Smothers comes into the scene, the car is modified by Vasek Polak into a roadster. Smothers sells the car and it goes to the Southeast where it runs BSR in SCCA races.

Don Bennett of Florida buys it and miraculously finds the original top and tail left over from the chop job. Folks do tend to save Porsche parts. The car comes back to California and into the capable hands of Kent Morgan. It's stripped to the bare chassis and everything is "replaced, renewed, or remanufactured." The engine was worn out, and went to Vern Covert (Porschop, Culver City, California) for rebirth. Kerry Morse (GTS Enterprises, Tustin, California) found original and reproduction parts enough to complete the job. Bob Hatchman (California Ideas, Arcadia, California), a high school buddy of Morgan's, did the unbelievable job of saving the body. The car now is literally better than new. When Porsche built it, it was a tool designed to do a job, not a shrine, not a grail. That status came later. Still, though the glasswork and paint are better by far than new, there is no silliness in the restoration, no chrome, no leather, no polished castings, no filing of welds. Warren Eads has done well to just leave it alone. Why did Morgan sell it? Why, to get his flat eight (Type 771) powered 907—the one with the 0.273 Cd. Of course.

I wish I could tell you that Miles and Patrick ran off from the big cars, but that only happens in Hollywood. But sometime during that hot afternoon someone idly asked who was leading the 2-liter class and unthinkingly I said, "Scooter—he's about 8th overall." The truth was, Miles was a lap ahead of him, so far among the big cars it was hard to remember that it was only a 2-liter. I have a magazine of the day that describes its staff as "The Visigoths of automotive journalism." They claim Miles was 6th overall. Would you believe a Visigoth? I remember Miles in that incredible Batmobile 3rd, and I like my version better. MT

SPECIFICATIONS

1966 Porsche 906-E

GENERAL	
Vehicle type	Mid engine, rear drive, 2-pass., 2-door coupe
Base price	$11,500 (1966)
ENGINE	
Type	Opposed 6, air cooled, 1991 cc (121.5 cu in.)
Induction	Port injection
Max. power (DIN)	220 hp @ 8100 rpm
Max. torque (DIN)	152 lb-ft @ 6400 rpm
Recommended fuel	100 RON
DRIVETRAIN	
Transaxle	5-sp. man.
CHASSIS	
Front suspension	Unequal-length A-arms, coil shocks and springs, anti-roll bar
Rear suspension	Dual trailing arms controlling hub, lower link and reversed upper A-arm, coil shocks and springs, anti-roll bar
Brakes, f/r	10.8-in. discs/11.2-in. drums
Steering type	Rack and pinion
Wheels, f/r	15 x 7.0 in./15 x 9.0 in.
Tires	Dunlop R-7 racing
DIMENSIONS	
Curb weight	619 kg (1366 lb)
Wheelbase	1485 mm (90.6 in.)
Overall width	1679 mm (66.1 in.)
Overall height	980 mm (38.6 in.)
Fuel capacity	100.0 L (26.4 gal)

PORSCHE 904 GTS.

CONCLUDED FROM PAGE 101

The black facia includes a complete range of instruments, and the various controls are easy to reach. As one expects, the short central lever operating the all-synchromesh five-speed gearbox is well placed, and extremely easy to use. The gear pattern is unfamiliar, but not enough to cause lasting confusion.

Driving the 904 does not present any special problems, simply because it is a very clear-cut car as well as a very forgiving one. The clutch, which requires a fair pressure on the pedal, is positive, with enough progression to give good feel to it. This quality, associated with an engine which is civilized by competition standards, allows you to make clean starts (I had by then sorted out the first gear) without either stalling or overdoing things. There are no problems with the 904 either in town or during acceleration runs. It can be driven in traffic, which is why a road version is being offered to those who want it.

When doing high speed tests, during which the car was eventually clocked at a very honourable 158 m.p.h., it remained extremely stable, with no wandering feeling even at maximum speed. Emergency stops in the wet from high speeds were reassuring as the car not only slowed down very quickly but it always did it without any trace of the unevenness which can occur with disc brakes in the rain. Those discs on the 904 seemed excellent, the only complaint being that you have to push too hard on the pedal. I did not encounter fade on my test but others who drove the same car earlier, reported some fade on fast, winding roads.

Curiously, our figures were generally a little better than those recorded by the factory. Perhaps my car had loosened-up by the time I tried it. These figures are very impressive and place this 2-litre car in the top of its class along with the 2-litre Abarth. Interestingly enough, the two are extremely close to each other, the Porsche having slightly better acceleration but the Abarth being about 3 m.p.h. faster. It compares well with cars in the bigger G.T. classes also. For example, it has quicker acceleration than the 3-litre GTO Ferrari which won the G.T. Championship last year. The 904 will accelerate from 0 to 50 m.p.h. in 4·5sec, to 60 m.p.h. in 5·6, to 80 m.p.h. in 8·4 and to 100 m.p.h. in 12·4sec, while the standing quarter-mile and kilometre were covered in, respectively, 13·7 and 24·5sec.

The gear ratios were well spaced with first taking one up to 47 m.p.h., second to 75, third to 102 and fourth to 130. Changes were fast and easy. The car seemed to be as much at ease on tight or fast turns, and its basically neutral steering makes it very easy to drive, especially since some slight oversteer appears as a warning when one is pushing it hard.

On the rapid Solitude course, which was slippery at the time, I was amazed to see how fast I could take it around corners. Considering the type of car it is, the ride was very acceptable. Noise inside the body was on the high side, as the engine is directly against the cockpit, ahead of the rear axle, and the exhaust is a very large, noisy one. Priced at about £2,860 at the factory, the 904 GTS is eye-catching, different and stimulating to drive.

Retrospect: 1966 Porsche 906-E

Photographed for *Motor Trend* by Neil Nissing Owned by Warren B. Eads

FIAT
4

Art & soul

One of the most charismatic – and certainly enigmatic – Porsches of the last 40 years, the 904 is now one of the most valuable, too. And having driven this superb six-cylinder car Peter Morgan (below right) can see why. Photos by John Colley

There's no denying that the 904 is one of the best-looking Porsches ever built, arguably even more attractive than any 911, and to many enthusiasts more gorgeous than any Italian supercar of the last 40 years.

Even so, somehow it's hard to get your head round the plain, simple truth that today even an average specimen – if there can be said to be such a thing – is worth at least £350,000. That's £25,000 more than a brand-new Carrera GT; four times more than a 996-model 911 Turbo.

Rarity has a lot to do with it, of course. Only around 100 904s were assembled, and that was fully 40 years ago now, in 1964. It's reckoned that most survive, but even so you've statistically more chance of finding a 959 for sale in the small ads. The vast majority were powered by Porsche's unique four-cylinder, four-camshaft Carrera engine that in race trim generated a lusty 180bhp (134kW).

In that first season, exactly 40 years ago, the 904 won the tough Targa Florio road race in Sicily, and would score countless other race and rally victories, too. As well as being strikingly beautiful the new car had secured itself a lasting place in the Porsche hall of fame.

But then in 1965 everything changed. Ferrari's new 206 Dino challenged the Porsche factory's racing team, and suddenly a great deal more was required of the little 904. Fortunately the then brand-new 911's six-cylinder powerplant provided the perfect answer, offering around 210bhp (157kW) in full-race trim. Thus was born both the Porsche 904/6 and a genuine automotive legend.

What makes these 904/6s so collectible today is that only 12 prototypes were built. And since they were all factory-run cars many either underwent very severe surgery of one form or another or else were crashed during a hectic 1965 racing season.

The programme of more or less constant modification came about because Porsche chose to take on the new Ferra i in the prestigious European Mountainclimb Championship. And for most of that 1965 season the Italian cars were demonstrably quicker. The series'

Carrera what?

During 1965 the 904/6 underwent frantic modification in order to reduce its weight and stiffen the pressed-steel, box-section chassis. The two objectives conflicted, though, and by the middle of the year it was obvious that a new car would be needed for the coming season.

Thus Porsche returned to the principle of the tubular-steel space-frame chassis for its next racing car – the Carrera 6 – and a body design that for the first time was developed entirely in a wind-tunnel.

The Carrera 6 (right) used an improved version of the type 901 six-cylinder engine, but with Bosch's new mechanical fuel injection instead of the previous Weber carburettors. Maximum power rose to 220bhp. In 1966 the new car would upstage the 904, winning the Targa Florio, and also capturing numerous class wins including the Daytona 24-Hours, the Sebring 12-Hours, and not least at Le Mans. ■

regulations limited the engine size to two litres, so the engineers were forced to try everything they knew to reduce the 904's weight.

That makes those coupés that survive in their original form very rare, indeed. It's why collectors are drawn to the 904/6 legend like moths to a candle. And why this still largely original car – chassis number 906-012 – draws admiring crowds literally wherever it goes. Put it this way. If an original, well-documented 904 with the four-cylinder, four-cam engine sells today for £350,000–£400,000, then a top-quality six-cylinder car with a good provenance will make around £450,000–£500,000.

The 904/6 is also one of that very small group whose value and position has evolved beyond that of a mere automobile. It has become almost a work of art, a symbol of both exquisite automobile design and of that all too brief moment in motorsport when racing cars were not only fearsomely effective, but also remarkably good to look at.

On the last occasion when I drove a 904 (see *Perfectly formed* on pages 88–94 of the August 2002 issue) I spent a long time simply enjoying the details of Ottocar Jakobs' superb example, in this instance one of the four-cylinder machines, and the manner in which the individual components came together to create a thoroughly effective racing car. And in that respect the 904/6 you see here is no different.

It's owned by Monaco-based Scotsman Irvine Laidlaw, and lovingly prepared by Martin Greaves of Classic Performance Engineering in Long Buckby, Northamptonshire. My experience of Ottocar Jakobs' 904 notwithstanding, I'll admit that to begin with I was quite concerned about the prospect of driving on the public road a car worth such a huge amount of money. Insurance is all very well, and we are scrupulously careful with anyone else's car, whatever it may be, but it would take only a moment's carelessness (by someone else) to damage – or potentially destroy – a piece of history.

Any 904 is incredibly small, for a start, making it quite difficult to spot from a distance if you're not looking for it. I'm fairly tall, admittedly, but it barely reaches my waist. In most such cars that would make getting in and out a real chore, but not here, where the tops of the doors roll over into the roof in order to make entry and exit that much easier. The devil is in the detail.

The extra space allows me to feed my legs over the high side sills and down into the footwell while supporting my weight on the roof and seat (but definitely not the flimsy door itself). I can easily imagine drivers doing it far more quickly than me during the Le Mans-style running starts that were so popular during the mid-1960s. (And probably not worrying too much about the door, either...)

Said doors (like the rest of the body shell) are formed from glassfibre, and have Plexiglas windows that simply slide up and down. There's no heavy winder mechanism, just a simple aluminium hoop attached to the bottom of the plastic that you drag down or up again. Even more ingenious is the fact that thanks to the friction between the plastic and the frame the window will stay precisely where you leave it, without rattling or gradually working its way down into the depths of the door shell. Detail again!

Massive 90-litre fuel tank (above) was essential for endurance racing. Weber 46 IDA carbs (right) give the six-cylinder engine a characteristic roar

There's a further pleasant surprise in the amount of space inside the cockpit, and for such a specialist machine the controls, both major and minor, are well placed. Sadly, the steering wheel is a modern-looking leather-rimmed job rather than the classic wood-rimmed Nardi item the car would have had back in the 1960s, but I can live with that.

Likewise the substantial roll-over cage isn't an original fitting, but rather an FIA requirement for modern Historic racing, and which proves that despite its value the owner of this car still uses it for its intended purpose. In fact, this is the third 904 cage Martin Greaves has built, and it perfectly suits the character of the car. Even the roof bars are sculpted to give more clearance to the driver's helmeted head.

As standard all 904s had fixed, non-adjustable seats moulded in to the floorpan, with the driving position set – somewhat laboriously, it has to be said – by moving the pedals and steering wheel as required. Luckily, though, I discover that as set for Irvine Laidlaw the driving position is just fine for me, too.

The deeply upholstered bucket seat offers a lot of support, and with the substantial four-point Willans harness buckled up I'm held firmly in position. As in many other machines of this size and style and agility it feels almost as if you're wearing the car rather than sitting in it.

Across in the passenger footwell I note another indication of how this car spends its summers these days – a modern Brantz trip-meter for road-rally events. The factory 904 decisively posted its international rally credentials in early 1965 when Eugen Böhringer and Rolf Wütherich finished second overall in the Monte Carlo Rally (see also *Jack of all trades* in Issue 46 of *911 & Porsche World*). Against all expectations the tiny glassfibre Porsches proved to be very competitive, and even today their pace and balance continue to worry all those purpose-built, short-wheelbase 911s.

Sensibly, Martin Greaves sits beside me for my first few miles in the car in order to explain the starting procedure and how it drives. Thanks to the pair of triple-choke Weber 46 IDA carburettors starting requires care to avoid

Racing type 901 engine (above) has twin-plug cylinder heads. Rear suspension (below) has anti-squat geometry

The factory view

Jürgen Barth started work at Porsche as an apprentice in April 1963. At that time his father, Edgar, was a Porsche factory driver with numerous international wins to his credit. Jürgen recalls being involved in building the 904s and watching his father drive the cars.

'As apprentices it was normal that we worked on the 904, but of course then I was doing only minor jobs. But I can remember watching my father at the Schauinsland hillclimb, and a little later I found myself helping three of our customers with their cars as well.

'My father raced the 904 to sixth place in the 1964 Targa Florio. In fact, he's in the car in the photograph on the cover of my book. But at the Nürburgring later that year he was driving chassis number 008, and had a bad accident after his brakes failed.'

Asked why the 911's then new six-cylinder engine wasn't in the 904 from the start, Barth has a simple explanation. 'The car was designed to take the six-cylinder, certainly, but *[in early 1964]* the racing version wasn't ready. Our engineers preferred to use the old four-cam because it had good power and it was reliable. The cars were tested with this four-cylinder engine, and also with the eight-cylinder, 2.2-litre engine. Later we used the set-up from both these tests for the six-cylinder.'

Not surprisingly, Barth's new book on the 904 (right) looks set to become the standard text on the car. For Barth himself, now a 40-year Porsche veteran, it's something he always promised himself that he would do. 'I have collected all the information about all the racing cars over my 40 years here at Porsche. And I mean everything, even when a customer sends me a letter.' (Barth has, of course, been involved with the customer sport department since the mid-1960s.)

'I filed all these letters by chassis number, and I think that in the future there is a danger that we may lose such information, particularly now that everything is routinely stored on computer. The new personnel don't know as much about these cars as the old guard.

'I decided about five years ago to write some books on the different Porsche types that I had worked with, and the 904 is the first one. There are books on the 906/910 and the 550 in preparation, too.'

Barth's book on the 904 is available only in German at present, but he hopes an English version (and French) will be available soon. ■

Modern steering wheel shows that this 40-year-old veteran is far from retired. Pedals (far right) and wheel adjustable. Wiper arrangement (below) pre-dates 993 by 30 years

Classic Performance Engineering

Martin Greaves set up Classic Performance Engineering six years ago after learning his craft with independent Jaguar specialist Pearsons near Silverstone.

He now works out of a busy industrial unit at Buckby Wharf, Northamptonshire, offering a full restoration service for classic and historic cars of all types. He also provides a race preparation and full trackside support service for owners of historic models.

The business is highly self-contained. Martin has three full-time staff and two part-timers. There's a well-equipped machine and fabrication shop alongside the general assembly areas. He has also established a trusted network of local suppliers for specialist work such as paint.

During our visit in March the workshop was a snapshot of 1950s' and 1960s' nostalgia, with two F1 Coopers, a gorgeous Elva-BMW sports racer, an AC Ace Bristol, a 'D'-type Jaguar and several Alfa Romeos undergoing work. A number of well-known collectors entrust Martin with the care of their entire fleets.

Irvine Laidlaw's 904/6 is a favourite, though. 'It's a lovely car,' Martin says with genuine enthusiasm. 'It's so adjustable, and it's from a period of racing that I love. But the best thing for me is that the owner wants it prepared absolutely right. That's the only way I could do it.'

Classic Performance Engineering is based at Buckby Wharf, Long Buckby, Northamptonshire; tel: 01327 844416. ■

flooding the engine. I let the fuel pump tick for a few moments, twist the ignition key to set the engine turning, and then ease in some throttle. With little complaint from the starter motor the flat-six comes to life behind me. And what a glorious noise!

Inside the cabin – with no sound insulation whatsoever – it's not unlike standing in the middle of the orchestra during a particularly spirited rendition of Tchaikovsky's *1812 Overture*. The crispness of the engine sound is superb, and gives me an immediate feeling of contact with the considerable amount of mechanical activity taking place behind my shoulder.

A few blips on the throttle to clear the spark plugs, and almost gingerly I try the five-speed type 904 gearbox. This was derived from the 901 production unit, but strengthened for racing use. The dog-leg first gear is selected by pressing down the shift lever with the palm of the hand and moving it left and back against a fairly strong spring. Reverse is to the left and forward, opposite first. It's far from ideal for road use, but at least there's synchromesh on all the forward gears – another typically Porsche detail.

At first I'm too gentle with the gear change, and Martin urges me to stop pussyfooting about. Even though the oil is warm I'm trying to ease the gears into mesh, as I would with the later 915 gearbox. But this isn't the way with this one. I have to develop a firm, almost indelicate, tug or push action, and get used to feeling the gear notch in at the end of the lever's travel. And, after a few crunches, I get the knack.

Pulling away is straightforward, and the clutch, although a modern Sachs racing item, isn't too abrupt (in fact, I stall only once!). But at low speed the engine, too, complains about my softly-softly approach. The spark plugs soot up again as I try to accelerate at half-throttle in second. The engine fluffs and coughs until finally it picks itself up off the floor again and we begin to fly. I'm learning that to drive this car fast you need to wring its neck – but skilfully.

So, with that open exhaust reverberating around the rolling hills of Northamptonshire, this wonderful car comes alive for me. With only around 650kg and 210bhp (119kW) to propel it the power-to-weight ratio (323bhp/tonne) is on a par with that of a modern 911 GT2. It would leave the 911 GT3 breathless in its wake. And, of course, the 904/6 is inherently more agile than those significantly heavier modern cars, no matter how good their steering, suspension and brakes.

Gaining confidence now, getting the hang of this delightful machine, I don't give the revs a chance to drop below 2500, and whenever I accelerate my foot goes straight to the floor. The upshifts are firm and decisive, while my downshifts – even if I say so myself – are straight from the soundtrack of John Frankenheimer's classic *Grand Prix* movie, with an evocative little yelp from the engine as I blip the throttle and snick the lever smartly through the gate.

So light is the car, in fact, that when Martin – trusting soul that he is – climbs out to let me go solo for a while it's immediately and notably more responsive to the throttle. The rev-counter needle flicks even more effortlessly round the dial, and for the first time I begin to enjoy the camminess of the diminutive flat-six. It doesn't really begin to go until the revs pass 4000, but then it comes all in a rush. Fantastic!

The 904's composure over bumpy asphalt is surprisingly good, and it reminds me how much we seem to have forgotten about ride comfort in modern sportscar design. Even on a mixture of 'wet' and 'dry' race springs at front and rear, respectively, with pencil-thin anti-roll bars, and not least old-fashioned oil-filled dampers, this 904 has a supple, compliant ride that would embarrass many later machines. It's easy to see why it was such a good rally performer.

In fact, there is some considerable sophistication in the suspension design. Besides the independent unequal-length-wishbone layout, the rear end has

Facts & figures

1964/5 Porsche 904/6

Body/chassis	Pressed-steel chassis with twin longitudinal box-section sidemembers and spot-welded crossmembers. Two-seat coupé body in glassfibre; 90-litre steel fuel tank
Engine	Type 901/20 air-cooled flat-six with magnesium-alloy crankcase and aluminium-alloy cylinder heads and barrels. Forged-aluminium pistons with titanium connecting-rods. Twin 46 IDA Weber triple-choke carburettors. Chain-driven single overhead camshaft per cylinder bank. Two sodium-cooled valves and two spark plugs per cylinder. Dry-sump lubrication system with separate 14-litre oil tank. Twin-coil ignition system
Capacity	1991cc
Compression ratio	10.3:1
Bore/stroke (mm)	80.00/66.00
Maximum power	210bhp (119kW) at 8000rpm
Maximum torque	196Nm (145lb/ft) at 6000rpm
Transmission	Type 904/1 five-speed transaxle with limited-slip differential
Suspension	*Front:* Unequal-length fabricated steel wishbones with anti-dive geometry and anti-roll bar. Coil springs with telescopic dampers. *Rear:* Reversed upper and lower unequal-length wishbones (with anti-squat geometry) and twin forward-facing radius rods; anti-roll bar
Brakes	Dual-circuit hydraulic system with Ate two-piston calipers and solid steel discs (made under licence from Dunlop)
Wheels & tyres	Pressed-steel wheels with aluminium-alloy rims all round and 185HR15 Dunlop racing tyres
	Front wheels: 5.5J x 15 inches
	Rear wheels: 6.0J or 7.0J x 15 inches
DIMENSIONS	
Weight	640kg
Length	4090mm
Height	1065mm
Wheelbase	2300mm
PERFORMANCE	
0–62mph	Not known (but four-cam car is 5.5 seconds
Maximum speed	165mph (see text)

anti-squat geometry built in to it. This helps to prevent the back of the car from sitting down under hard acceleration and at the same time introducing undesirable changes in the wheel cambers and toe angles. It was all pretty clever stuff for the mid-1960s.

During my brief drive I'm a long way from getting the tall Avon tyres warm enough to test the edges of this car's substantial performance envelope. But I can sense its tenacious grip. And with its mid-mounted engine the weight distribution holds no surprises, either. The car turns in to the few corners I tackle with real gusto with me barely having to think about it.

When I drove Ottocar Jakobs' four-cam car I was conscious of its superb agility. It was perfect for throwing around the sinuous mountain roads near the owner's Austrian home, and it was easy to see why a similar machine so decisively overcame Sicily's twists and turns to win the 1964 Targa Florio. But even during my short drive in this six-cylinder car I was conscious of its more long-legged feel. It's incredibly rewarding when you get it moving quickly.

The brakes, too, are something of a (pleasant) surprise. The 1960s' 911-derived discs certainly don't look much, but for this lightweight machine they are hugely effective. I wouldn't describe them as solid, but they have great feel. I can understand why Irvine Laidlaw uses his 904 for such varied competition. As well as being so much fun to race at all the major classic courses, it must be a real hoot to drive fast in road events such as the French Tour Auto.

Back in the 1960s the factory cars were good for 165mph at circuits like Le Mans. Today this one is usually geared for a maximum of 142mph at Spa-Francorchamps, and where better acceleration is required top gear is usually limited to around 125–135mph. That might not seem too remarkable compared to modern racing machinery, but on the Historic-racing scene the 904 is none the less capable of producing a few surprises, not least because of that impressive power-to-weight ratio.

This is Irvine Laidlaw's second 904. Prior to this car he owned chassis number 904-069. That was a quad-cam car, and brought him significant motorsport success before he sold it to

Shallow opening in car's nose feeds air to oil-cooler and front brakes. Door design (right) features exquisite detailing: note finger recess and wrap-over into roof area

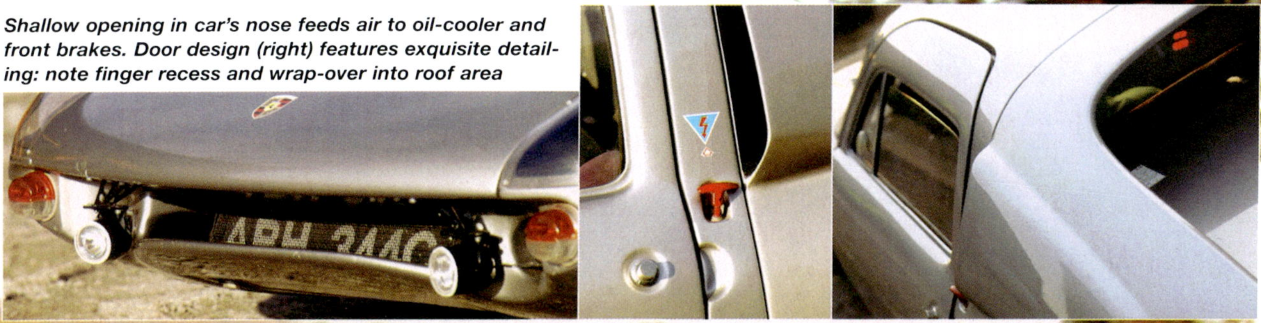

a new owner in Switzerland last year. In fact, the two cars overlapped for a time at Martin Greaves' workshop because the canny Laidlaw wanted to see how competitive the six-cylinder was before disposing of his four-cam car.

This second car had been involved in a collision on the Tour Auto before Laidlaw bought it, and had been repaired by a well-known German restorer. He brought the car to Martin Greaves for preparation for European Historic racing in early 2003. Martin and his team completely stripped the car and shipped the engine off for a full rebuild to German experts Knebel Motorsport in Siegen, near Stuttgart.

The rebuilt car was readied for Irvine and his co-driver, John Diffey, to run in the 2003 Gentlemen Drivers' Historic Championship. And the quality of Martin's preparation was proven when the car won no fewer than four out of the five rounds.

This year Irvine Laidlaw plans to compete in the FIA Historic sports-car series, with races at Monza, Spa and the Le Mans Classic, but he's also entering the 400km race at the Nürburgring. It's difficult to think of a more suitable machine to run in such an event.

Martin Greaves says Irvine keeps his cars only for as long as they are fun to own and drive. In which case I predict that he'll have this one for a few years yet. As well as its hugely competent all-round performance, this is a simply gorgeous machine to look at. With the front and rear sections open or closed it engages the eye, and draws a response from anyone with even the merest of enthusiasm for old cars. To see it is to want to touch it.

Even so, in spite of all its obvious qualities, explaining how a machine like the 904/6 can be worth up to £500,000 isn't an easy task. But looking past its huge value the mere fact that this car can still be found pounding the Tarmac – in anger, too – is a testimony to its basic appeal. It reveals a level of commitment far beyond mere speculation. ■

RSKapades!

In which Mike McCarthy, having found himself behind the wheel of an historic Porsche RSK at Bridgehampton circuit, delves into its history. Pictures by Mick Walsh

Mickey Mouse didn't take the lead straight away – Roy-boy had made it out of the chute first and was running front man for a while – but it wasn't long before he was there, with Taffy behind him, and they were one-two. Up and down the road through Berlin they zapped, hitting that high, high banking once a lap, squatting down as the G-forces took hold of them, pinning them down onto the brickwork. And then it happened: Mickey Mouse was out of the groove, swinging right up that banking, up, up and over the lip. A flash of silver, a blur, cartwheels, gone from vision, a crump and an explosion behind the massive earth pile. The crowds, stunned, just stared at open air where there had been, for a micro-second, a car, vertical, then – nothing.

No-one could survive that. *No-one.* All the same, Lorcher went looking. The crumpled, flaming mass was so distorted that no part could be singled out. Lorcher still looked, at the mass, all round it, further afield. Dear God! There he was, in among the bushes behind the earth wall! Lorcher rushed over. He was unconscious, but alive . . .

A strange, unearthly scenario, to be sure, but fact. At the Grand Prix of Berlin on September 16 1956, Richard von Frankenberg, the driver of a Porsche which he had nick-named 'Mickey Mouse', was leading the race from Taffy von Trips in another Porsche and Roy Salvadori in a Cooper-Climax when, for no reason that could be accounted for, the car suddenly veered up the steeply sloped banking that was such a feature of the race. The car flipped over the lip at the top, cartwheeling before it

Above: Murray 'The Legend' Smith at speed. Below: the incredibly complex engine, incredibly simple cockpit ...

thumped down into the car park beneath, its magnesium body instantly erupting into flames. Miraculously, von Frankenberg was thrown out as the car flipped and landed in the shrubbery, unconscious but little harmed.

The significance to this story lies in Mickey Mouse, more prosaically given the Porsche number Type 645 in-house. It was the prototype of the Porsche 718s, the models that turned Porsche from tiddlers to winners overall. Mickey Mouse's career was brief and, literally, meteoric: the 718's was long and meritorious.

Porsche had entered the sports-racing field (as distinct from the sports car field) with their Type 550, launched in 1953. It was a low-slung, lovely little beast, whose story was told in CLASSIC AND SPORTSCAR back in August 1983. It featured a ladder-type chassis, torsion bar suspension front and rear, and, initially, a developed version of Porsche's flat-four pushrod engine. In this form it was fairly successful, taking the 1½-litre class at Le Mans and the Carrera Panamericana for example –

RSK '005 ran with and without tail fins: some drivers loved 'em, others hated 'em. This is the form in which it raced at Le Mans in 1958, numbered 29, finishing third

hence the 'Carrera' Porsches. However, it wasn't long before a brand new power unit appeared, prosaically (as with all of Porsche's ventures) given a Type number – 547. Designed by Ernst Fuhrmann, who would later rise to head the company, it was a potent little unit but fearsomely complicated, incorporating no fewer than nine shafts and 14 bevel gears to drive the twin overhead camshafts on each bank of cylinders. It was blooded in the 1954 Mille Miglia when Hans Herrmann and mechanic (later driver) Herbert Linge took a 547-equipped 550 to a sensational sixth overall and first in class. As with all Porsches, its strength lay in its strength (it was perfect for long, arduous races such as Le Mans or the Carrera Panamericana) and performance.

Cockpit, with adjustable seat, is spacious and neat

In the mid-fifties the 1½-litre class was hot contested, often with names that are but a dim ar distant memory now – EMW and Borgward fro Germany, Maserati and OSCA from Italy, Lotus ar Cooper from this country. There was thus conside able competition for the Porsches, and in 1956 th 550A was introduced. The important changes we the replacement of the ladder chassis with a spac frame and a low pivot-point, swing axle rear suspe sion, while by now the Type 547 engine was giving healthy 130bhp. In a barn-storming year 550A finished fifth overall and first in class at Le Man but, much more sensationally, Umberto Maglic took a singleton to the Targa Florio – and won outright!

Oddly coloured panel on nose is actually a surface oil cooler. In the background is the derelict control tower for the Bridgehampton circuit, scene of former glories

However, that same year Porsche began to realise that perhaps their rugged little racers were just a mite *too* rugged. Lotus in particular had shown that lightness and superb roadholding could compensate to some extent for a lack of power. A slimming exercise was undertaken on the 550. The result was the Type 645 – 'Mickey Mouse'. Wheelbase and track were cut by several inches, the body was wrapped much more closely around the mechanicals, faired-in headlights replaced the standard units, and there were wishbones at the back at last.

The reduced frontal area and lighter weight – body panelling was in magnesium – did indeed work but only in a straight line: maximum speed rose from 158mph for the 550A to 162mph, but the handling was, to put it mildly, treacherous, for it would switch from understeer to oversteer as quick as blink. Richard von Frankenberg nick-named it 'Mickey Mouse' because of this. Its first race was at the Solitude in July 1956 where various ailments befell it, though it eventually finished fourth. Its next, and last, was at that infamous Berlin Grand Prix.

'Mickey Mouse', though, had taught Porsche a lesson: a lower frontal area could increase speed. Over the winter of 1956/57 a new prototype was built up, and given the Type number 718. It drew heavily on 550A practice, whose wheelbase and track it retained, and indeed aft of the cockpit there was little change. Up front, however, the tranverse torsion bar/trailing arm set-up was modified. The lower tranverse arm stayed as was, but the upper pivot-points were raised, and the transverse torsion bars, instead of spearing straight across the nose, were angled down to the centre line of the car. Viewed from the front it looked like the letter 'K' laid on its side, and the new model was nick-named the RSK. The system, designed to give some beneficial camber changes to the front wheels, didn't last very long – but the name did. RSK, 718, it didn't matter what you called the car – it went like a dingbat.

Part of the reason for this was a further increase in power via an increase in capacity. In 1957 the European Hill-climb Championship had been revived, with a maximum capacity of 2-litres, basically to suit the Italians. Porsche looked at it and reckoned they were in with a chance, so produced two more versions of the 547 engine, the 547/4 with an 87.5mm stroke (standard wear was 85mm) and the 547/5 with a 90mm bore: capacities were 1587cc and 1678cc respectively, and outputs were 150bhp and 7200rpm and 170bhp at 7800rpm. On the whole, from 1958 on, the works cars ran 1.6s, privateers took over the 1.5-litre class, and the 1700 was hardly ever used.

Other changes included an overall lowering by some 4ins thanks to sleeker bodywork, a full-width screen to meet new regulations, ducts right in the tail to administer cooling air into (not as would seem out of) the engine bay, and an oil cooler

An RSK only put in two appearances in 1957, once for practice at the Nürburgring in May, when it didn't race, and once at Le Mans when it did. For the latter, a pair of squared fins were added to the tail to give the dart some feathers. Whether they worked or not seems to be a moot point: some drivers hated them, others wouldn't climb into an RSK without them. In practice for Le Mans, in fact, the RSK survived an enormous spin while it was being driven by Edgar Barth, which rather indicated that they might work in a narrow angle of travel, but once beyond that and – whoops . . . In the race the car went out with accident damage.

Behra and Herrmann were in contention for the whole of the race, never dropping below sixth, and it wouldn't have taken much to win

However, the RSK was obviously the way to go, so more development work was carried out over the winter of 1957/58, which was when the car featured here, chassis number 718005, was built. Gone for a start was the K front suspension, replaced by the standard item. At the back a pair of Watts linkages, placed fore-and-aft and centred on the hubs, reduced unwanted toe-in – and hence steering – effects. And – heresy – coil springs replaced torsion bars. The definitive RSK was up and raring to go.

And go, indeed, it did, with '005 having its share of successes. At Sebring in March 1958 feisty Jean Behra and Edgar Barth driving an RSK 1600 harried the big 'uns before an oil seal forced a retirement.

It rained hard at Le Mans in 1958, but that didn't stop Behra and Herrmann (here driving 'our' car, number 29) to a magnificent third overall, with number 31 fourth

Behra takes over number 29 during a night fuel halt

Behra drove a 1500 RSK in the Targa Florio that year, and brought it home a fighting second – *Autocar* reported that 'Behra, due to his starting order, took the flag and the ovation in the pits was terrific – his was the moral victory and the Italians loved it'. The winner was a 3-litre Ferrari driven by Musso and Gendebien.

At the Nürburgring Porsche stuck to the 1500cc class again, finishing sixth and seventh overall and in the shorter 500km race at the same circuit Behra once again won overall in '005. The entry list for Le Mans comprised two 1500s for Barth and Paul Frère and the burly Dutchman Carel de Beaufort teamed with Herbert Linge, plus a 1600 entrusted to Behra and Hans Herrmann. The latter was chassis number – yes, you've guessed it, '005 . . .

And, in fact, the Franco-German pair were up in contention for almost the whole race, never dropping below sixth. It was a dreadful year weather-wise, with almost constant rain, and inches deep puddles in places. The Porsches buzzed round reliably – *Autocar* called them 'wonderfully constant and rapid' – though it would seem with some circumspection, for the same magazine also reported that they 'seemed to have remarkable adhesion in the wet, and their tails to snake less than in the dry'. It's wonderful how a careful right foot can keep a car in a straight line . . . So well did the cars run, though, that *Autocar* predicted at one point that 'It was already clear that it would require no exceptional stroke of fate to place a Porsche in the lead of this desperately gruelling race', but in fact all was not perfect for at one point '005 lost its left front brake, requiring a new drum and shoe, but at the end of the 24 hours she was still there in third position, behind the Hill/Gendebien Ferrari and the Whitehead's Aston Martin. What's more, Barth and Frere were fourth, and de Beaufort and Linge fifth: for the finish the three cars arrived at the post in line astern . . .

As Karl Ludvigsen wrote in *Porsche – Excellence was Expected*, 'This outstanding performance won the attention of the racing world, confirming the ability of these small, tough cars to challenge for victories in the great classic races'.

For the Avus race, the one in which von Frankenberg had crashed so mightily in 1956, the Porsche team included Behra in '005 *sans* fins, curiously enough, and he won by 0.8sec.

Once again Porsche contested the European Hill-climb Championship in 1958, though by now the capacity limit had been reduced to 1.5-litres – perfect for Porsche, of course. However, Ferrari and Maserati withdrew, leaving the field to the Stuttgarters in the form of Taffy von Trips (who invariably used '005) and Behra, with opposition from Borgward. Taffy, by dint of wins at Mount Parnassus in Greece, Monte Bondone in Italy, and Gaisberg in Austria, plus a second at Mont Ventoux in France, took the Championship from the Borgwards of Herrmann and Bonnier. Jean Behra was so incensed by all this that he suggested that von Trips' older car, '005, was quicker than his 1958 machine. They held a grudge race at Zeltweg where they swopped cars – and von Trips won by less than second in the new car.

For 1959, Porsche carried on with the successful RSK formula, though the latest works cars had yet another revised rear suspension, this time by pure wishbones and coils. That more wasn't modified can be partially pinned down to the fact that Porsche were getting into single-seater formula.

The FIA had announced that, from 1961, F1 would be for cars with a maximum capacity of 1.5-litres (Porsche territory), a formula that, up until then, had covered F2 cars. In fact, an RSK, converted to a single central seater (but with RSK fully enveloping bodywork) had been entered for the F2 event at Rheims in 1958 for Behra – and he won. For 1959 there were true single seaters on the stocks, and the works RSKs suffered to some extent. Behra had transferred his allegiance to Ferrari for the World Championship events, but bought a customer RSK for the 1.5-litre races, while Bonnier came over from Borgward. That the new wishbone rear suspension worked was proved by von Trips and Bonnier coming home an easy third at Sebring early in the year, its softer rear springs allowing the power to be put down much earlier in

On its way to a Targa victory – '005 at the start, 1959

the corners, though at the expense of the inner front wheel waggling about in the air. 'Our' car, '005, came home fifth, driven by John Fitch and Edgar Barth.

Thence to the Targa Florio. Two of the new wishbone cars were entered for von Trips and Bonnier and veterans Maglioli and Herrmann, both 1.6s, backed up by '005 with a 1.5-litre engine for Edgar Barth and Wolfgang Seidel. Facing them were the official Ferraris with the likes of Behra, teamed with Tony Brooks, Cliff Allison, Dan Gurney, Olivier Gendebien and Phil Hill. One by one the Ferraris dropped out – Behra overturning his car, but managing to right it again and trundle round to the pits for Tony Brooks to take over – and by lap four out of 14 Bonnier and von Trips were out in front, all Ferraris gone. Bonnier tramped on mightily, until at half-distance the car was 14 minutes in the lead – from Barth and Seidel in '005. And then, at Collesano, with less than 25 miles to go, on the last lap, von Trips ground to a stop, his gearbox gone. Barth and Seidel crossed the line first in '005.

The rest of the year was pretty dismal though, especially at Le Mans were *all* the works cars retired. A hot cam had been tried and proved to be just too hot . . . Then, at the Avus, von Trips won the wall-of-death race, but this time the name proved accurate: Behra, in his private RSK, emulated von Frankenberg and flipped over the banking. He was not so lucky, though, and was killed instantly when the car hit a concrete post.

By the end of the year, just before the Tourist Trophy, Porsche were in third place with 15 points for the Manufacturers' Championship, behind Ferrari on 18, and Aston Martin on 16. In the race itself the leading Aston caught fire in the pits and the Bonnier/von Trips RSK moved into the lead! The second Aston was called in and Stirling Moss put on one of his brilliant displays to get ahead once more. Ferrari were as chaotic as usual, and misinformed Tony Brooks, who was in third place, as to his real position: when he did speed up it was to late, and the Porsche finished ahead of the Ferrari by a few seconds to give Porsche a second overall – and second in the Manufacturers' Championship!

With the withdrawal of Borgward from racing in 1959, too, the Hill-climb Championship was a walk-over for Porsche. Edgar Barth used '005 to win five overall firsts and win the Championship fairly easily.

For 1960 the FIA decreed new rules which included such absurdities as a much larger windscreen (which would, of course, have a disproportionate adverse effect on smaller engined cars) and luggage space. Porsche responded with the RS60 – but that is yet another story in the 718 saga, as is the F2 car, which will have to await another time.

Driving Impressions

Murray (The Legend) Smith is a trusting soul. I mean, I've driven a wide variety of cars, but always on the road. I leave the track stuff to our man Green, who's happier sideways. Me? I'll never be a World Champion, so why pretend? But then along comes Murray, and offers to let me drive his RSK – a very special RSK, as we've seen – at Bridgehampton on Long Island, near New York, New York, in the USA at that. Who could resist?

Murray (an expatriate Brit, by the way, who used to race in the JCB series here) has owned a couple of Porsches, including the Behra single-seater F2 device which he sold to buy an Aston DBR4. He still hankered after a Porsche, though, and found one sitting literally in a barn he owned, where it was being stored for its owner. He was taken by its history (it's one of only nine works RSKs), in particular its third at Le Mans, so he bought it. The car was sitting on its bottom, scruffy, but with a new, albeit cosmetic only, engine in its rump.

The engine went to Porsche twin-cam expert Jim Wellington in California to be turned into a real 'ricer', while Tony Dutton (another ex-pat) and Bob Linker of Northumberland Engineering coped with the rest of the car. And a damn fine job they've done too: Murray insisted on as many of the original parts as possible being retained, and they were. Look closely and you'll see the little lips over the wheel arches: look back to pictures of the car in 1958 and they're still there. The same ones ... What's more, Murray calls it a 'European' restoration, not an 'American' one: it's for show as much as go, and it goes as well as it looks. Which is sensational.

Bridgehampton hasn't been used for some years now, but it's all there and in reasonable repair – the only problem areas are where sand from the surrounding dunes has drifted across the corners. There's a longish straight to start, an oh my gawd dropping right-hander, then the most amazing series of lefts, rights, lefts, right, up, downs, ups, downs you've ever seen. It takes about 100 laps simply to learn it! And it feels so narrow too: you think of Cobras bombing through and you think 'Gulp!'

Of course, I was nervous before I went out. So would you be. This was a real racer, all blood and thunder, not a namby-pamby road car dressed up.

I needn't have worried. There were two things that made you realise this was no road car, but the rest was a doddle. It's even easy to climb into and out of, and the latter was done with considerable regret, I might tell you ...

The two problems? The first concerns the gear-change. First gear is for starting only, and is a most peculiar round-the-corner dog leg, but once you're on the move this, of course, become irrelevant. The major problem comes in with the fact that this has to be one of the vaguest changes I've ever used – you can waggle the lever round and you feel lucky at first if you've found *any* gear. It desperately needs a gate to guide the lever, but that wouldn't be original. Murray swears you get used to it eventually, and I'll take his word for it, but I had hassles ... Mind you, find a gear and snick! It's in instantly.

The other problem concerns the power delivery, and this proves I'll never be a 'ricer'. Below about 4500rpm there's ... I was going to say nothing, but it's worse than that. Try feeding in even a mite, a tittle, a jot of throttle and that complex little engine takes considerable exception, coughing and spitting and popping and generally behaving in a very foul-tempered way indeed.

Two problems and a bare couple of hundred yards covered. Not a good omen. And then, suddenly, it's at 5000rpm, and Zap! Wham! Kapow! Wow!

You'd never believe that a 1600 could pack so much wallop. I'd been given a 7000rpm red line (Murray takes it to 7600rpm 'in an emergency!') and from 5000 to 7000 it's the same sensation as a turbocharger coming in. Now I know, really *know*, what they mean when they say an engine comes 'on-cam'. Now I also see why real drivers raved about this unit. For a start, it feels much more like a really punchy 3-litre, not a puny little 1.6-litre. For another, it even *sounds* likes a 3-litre, with a surprisingly deep and gruff (and loud) exhaust bark. You simply can't believe a car this small can go so quickly or make so much noise (Murray reckons maximum on current gearing is about 158mph, and the 0-60mph time around the 5secs or less region – that is flying by anybody's reckoning).

But camminess and gearchanging are the only problems, and then they're only problems to a novice like me. The clutch is gentle and light and smooth – you could use it in traffic without a qualm – and the steering simply sensational. Light, accurate, it's full of feel and with absolutely no kick-back or fight or ructions of any sort: again, your granny would love it. Brakes are heavy at first, but when you have to use them a hard shove shows that they give nothing away to discs.

I wasn't about to bend The Legend's car – it was off the next day to Leguna Seca – so I didn't try its cornering powers to the limit. They were, in any case, far above my own. So, driving it at – what, six tenths? – all I can tell you is that this little beastie corners as if on rails. Mild understeer can be turned into neutrality and then mild oversteer with judicious use of the throttle, all very predictable and comforting. Believe you me, there is nothing in my experience like powering a Porsche RSK out of a tight corner, the engine on cam.

But the biggest surprise has to be the comfort levels. There's even a set of slider rails for the seat, so there's no ponsing about with cushions of foam padding. All the controls are perfectly placed, right where they should be, the pedals are fine for heel-and-toeing, but above all the ride would put many a modern GTi to shame. It did, in fact, for it was considerably better than the Dodge Shelby Charger which we were using as a camera car.

Mind you, on the limit it seems to be another story. Ken Miles, an RSK driver of considerable talent, had this to say about it in *Sports Car Graphic*: '(It's a car) that is exceedingly brittle to drive: whatever happens happens fast. If the engine rpms are allowed to fall too low as you enter a turn, opening the throttle produces nothing but a vague gasping sound. The engine splutters and coughs until, just as you have given up hope, it takes hold with a mighty surge of power then ... watch out! There goes the tail! If the RS was a taut car compared to the sloppy old 550, the K is tuned to piano wire tension ... On a slick, low speed turn the car understeers horribly ... at over 100mph the understeer resolves rapidly into an oversteer condition which must be corrected with a very gently movement of the steering wheel; any sudden or violent movement throws the car out of control!'

Murray Smith partially agrees: "Driven gently, it's absolutely forgiving, but on the limit ... The problem is the transition period, the turn-in. Once you're in the corner, and the car's set itself up, with the swing axles out, then it's all right – unless you lift off or try and change in the middle of a corner. It'll snap round just like *that*. The later wishbone cars were much better in this respect.

"But doesn't it feel lovely and taut? And doesn't it have a watch-like precision? (*Agreed – MM*). And think of all those people who've raced it - Behra, von Trips, Bonnier, Barth, Seidel, Fitch – it's living history!"

And so say all of us ...

This has to be someone terribly famous, doesn't it?

NEW MODELS

PORSCHE GT 904 released

SOME six months ago Ferry Porsche explained in an interview with the *Christophorus* that a new Grand Touring car was being designed and was in development for the 1964 motor sport season. The preparations have, after extensive tests and endurance runs, made such progress that this new car can now be presented to the public.

First of all, this new vehicle will draw the attention of those sports drivers who are really interested in active participation.

At Porsche there always has been a production car specifically designed for the competition minded sportsman in which at least a class victory could be assured and often an overall win. Thousands of international successes prove the quality of these cars which were sold to the public. These cars, destined to be run in motor sports events, were built some years ago in the Spyder version, RS and RSK and finally as Grand Touring cars with 1,600 c.c. and 2,000 c.c. engines.

Low and bullet-like lines in front.

At rear, a touch of the Italian.

Today there comes the Porsche Type 904.

A light but rigid box-section frame carries the fibreglass body, which acquired its smooth lines from the experience gained in last year's prototype races.

Front and rear suspension consist of A-frame links and coil springs similar to the Porsche Grand Prix cars. Of course, all four wheels are equipped with Porsche-ATE dual circuit disc brakes. As in the past, 15 in. wheels are being used since Porsche does not intend to build "airport-racers" but rather cars which can be entered successfully in all sorts of races including a Targa Florio. In addition, the owner of such a car must have a chance to compete as well in a rally on all kinds of roads as well as drive his car to the races he enters.

The very reliable 2 litre 4 cylinder engine — with which the Grand Touring World Championship up to 2,000 c.c. was won twice — will be used in the new G.T. The engine now develops 180 b.h.p. and lies in front of the rear axle. The car is equipped with a synchronized five-speed transmission having specially selected ratios. The concept which has proven to be utmost in the design of the Porsche Grand Prix cars served as the basis for the construction of the Porsche Type 904. This car, which is 150 kg. lighter than the previous G.T. and has an improved chassis, has an extremely low frontal area with a height of only 106 cm. which gives a very aerodynamic form. With all this, the driver is equipped with a very potent instrument.

The surprisingly low price of 29,700 DM, in comparison with other competition vehicles, will be another point that will make this car more than just an item for the privileged few who can afford the best at any price. The interest already shown by private drivers ensures that quite a number of these new cars will represent the German colours during the next motor sport season.

Left front suspension detail.

The stripped frame has massive members.

TECHNICAL DATA

Engine

Type	Four stroke cycle, horizontally opposed cylinders, overhead cams air cooling.
Bore	92 mm.
Stroke	74 mm.
Displacement ...	1,966 c.c.
Compression ratio	9·8 : 1
Performance (DIN)	180 b.h.p. at 7,000 r.p.m.
Torque max. ...	20·5 mkg. at 5,000 r.p.m.
Carburettor ...	2 Weber dual throat downdraught
Lubrication ...	Dry sump

Frame:

Front suspension...	Transverse double A-frames
Front springs ...	Coil with coaxial shock absorber
Rear suspension ...	Transverse double A-frames
Rear springs ...	Coil with coaxial shock absorber

Dimensions:

Overall length ...	4,090 mm.
Width	1,540 mm.
Height	1,065 mm.
Wheelbase ...	2,300 mm.
Track, front ...	1,314 mm.
Track, rear ...	1,312 mm.
Ground clearance	120 mm.
Fuel tank	110 litre (29 gal. U.S.)

courtesy of Eugen Böhringer

AN IDEA THAT SNOWBALLED

TEN YEARS BEFORE THE LANCIA STRATOS SCORED ITS FIRST MONTE CARLO WIN, A MID-ENGINED SUPERCAR BLASTED THROUGH THE ALPS. HARTMUT LEHBRINK EXPLAINS HOW A PURE PORSCHE GT RACER ALMOST SPOILED THE MINI'S PARTY

IT WAS SPUR OF THE MOMENT – AND IT had a tinge of the impossible about it. But it was to bear fruit.

The *Gran Premio Internacional de Turismo*, Argentina's gruelling saloon car marathon held on open roads – and won for a second straight year by Eugen Böhringer aboard a works Mercedes-Benz 300SE – had been postponed a week because of torrential rains in the San Juan region that had washed away or flooded parts of the route. This meant that the competitors did not return to Buenos Aires until November 7, whereupon Karl Kling, the former GP driver who was Merc's team manager, decided that it was too late to enter his cars for the 1965 Monte Carlo Rally: his equipment had still to be shipped back to Europe lock, stock and barrel, and his team had lost a crucial eight days. There was another reason not to rush back: the Monte Carlo's regulations, as well as most of its sections and stages, favoured small and nimble front-wheel-drive cars such as the Saabs and Minis…

But Böhringer, still addicted to competition at the age of 42, had a brainwave. He telexed Porsche's sporting director, Baron Huschke von Hanstein, to ask if he was prepared to entrust him with a 2-litre 904 GTS for the upcoming event. The new model had been rallied with some success by privateers in Spain, France and Belgium in 1964, and had shown well in the Targa Florio and Coupe des Alpes, but a long-distance mountain rally in *winter* was something far removed from the car's original remit. The moustachioed baron's answer was succinct: "You must be off your nut!" Böhringer, though, would not be denied. He telexed back: "You're off your onion yourself. We can work wonders with that car."

Back in Germany a fortnight later, Böhringer called on von Hanstein and, blessed with the gift of the gab like nobody else, finally coaxed him into giving the project his thumbs-up. "What I need from you is the car itself, works service, the tyres and the co-driver," Eugen explained.

As to the latter, he opted for Rolf Wütherich, the Porsche mechanic who had been seriously injured in the road crash that killed James Dean in September 1955. Wütherich knew Böhringer well, having built a Formula Junior fitted with a Mitter two-stroke within earshot of Eugen's hotel in Rotenberg, a Stuttgart suburb. Rolf had a knack for technical things, but tended to stutter when excited – hardly ideal pace note material.

Above and left: 904GTS proved adept at dealing with snowy conditions. Below: Böhringer's local motor club helped out; basking in interior heat

all: courtesy of Graham Robson

As the Monte project took shape, von Hanstein warmed to the idea and optimism soon prevailed among the Porsche tycoons: with Böhringer at the wheel of their spectacular coupé, winning the world's most famous rally was a distinct possibility. There was, however, a potential obstacle that had to be overcome as quickly as possible. Would Böhringer's Mercedes-Benz employers consent?

'The car's road-holding was fantastic, not least because of it's perfect weight distribution'

At least nothing formal stood in the way. "I never had a written contract [with Mercedes]," explains Böhringer. "Driving for them was just a matter of a telephone call or a friendly little chat over dinner, capped by shaking on the deal. No red tape at all."

Prof Fritz Nallinger, the Mercedes director in charge of all the marque's racing activities, was on holiday in St Moritz when the request was made, but he generously gave the go-ahead at once, ringing Ferry Porsche with the good news himself.

Once the basic decision had been taken, the Porsche racing department acted quickly and efficiently. "With such short communication channels, the Porsche people performed in one day what would have taken their Untertürkheim colleagues a week," muses Böhringer.

Eugen was given a stock 904 for a week to familiarise himself with a car which he had never driven before, and which was very different from the big Merc saloons he was used to. The first snow had fallen in the Black Forest and he plunged in, possibly the first time a 904 had hit icy roads.

In the meantime, his Monte Carlo mount – chassis 006, a veteran of Targa Florio, Le Mans and the Tour de France – was being meticulously prepared. Its ground clearance was raised and an aluminium undertray fitted to cope with the snow and bumps. The passenger seat required by racing rules was just a dummy, so it had to be replaced by a proper one, although Wütherich would still have to live with the battery mounted in his footwell. Both the steering wheel and pedal set-up were adjustable and Böhringer quickly found a suitable long-distance driving position.

The racing 904s had no heater or demister, of course, so a petrol device – developed by Porsche in conjunction with Eberspächer during WWII to heat the engine compartments of tanks – was installed. It proved almost too effective. "We never needed to wear a pullover during the rally," laughs Böhringer. "But the problem was that the thing could not be regulated, just switched off or on, the noise of which always frightened the officials at the checkpoints."

The shortest of four potential rear axle ratios was chosen, reducing the car's top speed to 230km/h. At the same time second gear was made a bit longer and third a little shorter. This meant that the drivetrain was almost identical to the Targa Florio set-up used by the 1964 winners, Baron Antonio Pucci and Colin Davis. Böhringer was told that he could rev the noisy four-cylinder to over 7000rpm, but wisely decided not to exceed 6500 and instead make best use of the good torque from the Ernst Fuhrmann-devised unit.

The tyre situation was overseen by Porsche engineer Peter Falk, who was also going to take part in the rally as Herbert Linge's co-driver in a pre-production 911. It was a complex situation: Böhringer had the choice of various makes – Dunlop, Metzeler and the excellent Finnish Hakkapelita rubber – with either 125, 250 or 500 spikes of differing lengths shot into every tyre to cope with the varying conditions that were typical of this event. As Porsche could only man a couple of depots, Böhringer's local Motorsport-Club Stuttgart came to his aid, its enthusiastic members swarming all over the Alpine sections in particular.

Serious practice began soon after the turn of the year, von Hanstein providing Böhringer with yet another 904. "The car's road-holding was absolutely fantastic, not least because of its perfect weight distribution," ➸

Arriving in Monaco. Sadly for Böhringer, the end-of-rally blast around the track had been shelved for 1965

LAT

marvels Böhringer. He found it to be very user-friendly. His inexperience caused him some spins, but the car was never unduly "vicious", and soon he was throwing it accurately into tight hairpins, a useful skill given that its turning circle was a commodious 13 metres.

Böhringer was also full of praise for the braking of this 700kg lightweight: "We would arrive at a corner, say 50km/h faster than in the preceding year with the big Mercedes, but the braking point was the same."

Everything looked good, especially as the route was deemed to be faster and easier than ever before. Böhringer was confident of success. And then the weather turned with a vengeance. Epic snowstorms and high winds in the Massif Central ensured that this was one of the toughest Montes ever. The 904 was surely out of its depth...

Having started in Bad Homburg to reduce the overall distance to a minimum (unlike most others, who had chosen Stockholm, London, Warsaw and even Minsk), Böhringer's 904 was one of the last – only four cars seeded at 150 and above made it through – to survive the Arctic conditions that sidelined the bulk of the field, including the works sister 904GTS of Pauli Toivonen.

Peering out through that steeply raked windscreen, Böhringer drove superbly to set the fastest time over the snow-swept St Apollinaire stage... but he had still clocked up four precious minutes of road penalties – the lifeblood of the event – on the run to Chambéry and the struggle over the first two stages, Col du Granier and Chamrousse. He clocked the fastest time over the shorter, much easier Levens stage, and arrived in Monaco in fourth place. He would have been third but for the 20 penalty points collected for an overly noisy exhaust. Ahead of him were the Citroën DSs of Lucien Bianchi and Bob Neyret, and the leading Mini Cooper S of Timo Mäkinen, the only driver still without penalty.

After a day's rest in Monaco, just 35 cars – from the 237 that had started the event – took up the challenge of the last night, two 300km loops through deep snow and hard-bitten ice. Böhringer was just pipped on the opening Gorges de Piaon stage, and thereafter chased Mäkinen, the big Finn setting fastest time on the remaining five tests. Timo's night was not without drama, however, and a broken contact-breaker near Sospel cost him 4min. This meant that only Böhringer and Erik Carlsson's Saab 'cleaned' the section. The latter, however, was already out of contention because of iced-up carbs on the run down to Monaco; but the former, most definitely was in contention, and moved up to second after Bianchi had crashed out on the wrong tyres.

Mäkinen had dominated the event, but Böhringer had almost stolen the show. Von Hanstein was quietly content but would definitely have preferred first place. When Wütherich, who had shouted himself hoarse in his hopeless contest against the raucous power plant, groaned, "Huschke, water" at a checkpoint, the aristocrat ignored his plea. "Herr von Hanstein, water," repeated Wütherich.

"But your engine is air-cooled," came the unhelpful answer – eventually. At which point Wütherich had to grab a handful of snow as Eugen roared away. Von Hanstein had been indefatigable and ubiquitous in his efforts. He expected the same from his men. And more.

SUPERB MINI VERSUS SUPERCAR

Chambery-Monaco	SS1	SS2	SS3	SS4	SS5	Total
Timo Mäkinen	41m 28s	**30.25**	20.32	**12.47**	6.59	**112.11**
Eugen Böhringer	52m 55s	35.05	**19.55**	13.25	**6.20**	127.40

Stages: SS1 Col du Granier (47km); SS2 Chamrousse (35.5km); SS3 St Apollinaire (17.5km); SS4 Col de Leques (14km); SS5 Levens (7km)

Monaco-Monaco	SS6	SS7	SS8	SS9	SS10	SS11	Total
Mäkinen	6.55	**24.39**	**22.29**	**25.10**	**22.51**	**24.28**	**126.32**
Böhringer	6.32	24.47	23.44	26.35	23.15	26.07	131.00

Stages: SS1 Gorges de Piaon (6km); SS2 Col de Turini, east-west (26km); SS3 Col de la Couillole (21km); SS4 Col de Turini, west-east (26km); SS5 Col de la Couillole (21km); SS6 Col de Turini, west-east (26km)

NB: Times in bold indicate a fastest stage time

Stage winners: SS1 Erik Carlsson (Saab), 40.00; SS6 Peter Harper (Sunbeam), 6.29

Continental Correspondent — JERRY EVERETT—Goes driving in a new production sports-racing car.

160 MPH PORSCHE 904 GTS

THERE have been Porches before this which would touch an honest 160 mph — but few of them were made in FIA-approved series of 100 or more for sale to anybody with a (relatively) modest amount of money and an urge to go GT racing.

The new 904 GTS moves the Stuttgart house into that category for an ex-works price of roughly 7425 dollars. For comparison, that is something like the price of two of the cheapest 75 hp Porsche road cars or seven and a half 1200 Volkswagens, on the German market.

For another thing Porsche deserves full marks for actually putting the production GT version of last season's prototype on the open market for anybody with money and a modicum of sense. Further, the factory hopes and expects to find enough privateers to handle the 180 hp 904 so that they won't have to race a works team. Using the same body, Porsche factory drivers will sit ahead of a flat-eight engine instead and remain in the prototype class for such classics as Le Mans.

Let's break the 904 (with engine designation 587/3) down to its components — some almost vintage Porsche, others startlingly new for that small specialist factory.

They decided, in the face of all rumors, to remain true to the twin-ohc, ex-Carrera, four-cylinder engine for 1964, but made various changes to achieve their 180 DIN horses. The valves are larger and the camshaft timing is fairly radical to say the least; 97/98 for intake, 88/60 for exhaust. Pistons are altered and the valves now run 49 mm in diameter on the intake side, 43 mm on the exhaust.

Further motor figures, since we're on the subject, include a relatively staid compression ratio of 9.8:1, while maximum horsepower comes in at 7000 rpm. Maximum torque of 148 ft./lbs. is reached at 5000 rpm. They fit two Weber 46 dual downdraft carburettors and the bore x

The new Porsche is built purely for speed. The single arm windscreen wiper seems ungainly but works quite efficiently.

If you are thinking of competing at Le Mans, or having a bash at the Targa Florio — here's your chance. All you need is about £4000 and a driving licence.

The sleek GTS can be bought over the counter and could even be used for day to day transport if you don't worry about luggace space.

stroke runs 92 x 74 for 1966 cc. The lubrication is on the dry sump pattern.

A major change as far as Porsche road cars are concerned is the mounting of this powerplant (or the six or eight when they so choose) ahead of the rear half-shafts in the Spyder manner. The fact that there is no luggage space as a result rather underlines the non-road specification of the car, despite Porsche claims that it would be good for the Targa Florio or rallying. I'd hate to tackle the forest paths of the RAC Rally or the gravel crowns of the Liege with an absolute maximum of just under 5in. road clearance.

The three cars Porsche provided on debut day, one to photograph and two to ride in, with either European hill climb champion Barth. or German GT champion Herbert Linge driving, had nothing like the prescribed 12 centimetres of free air under their lowest part. We asked what they intended to do for Le Mans, where the scrutineers have a box they shove underneath to check such things.

The answer is that customer cars — the four-cylinder production models in other words — will be delivered with Dunlop SP tyres and they furnish the clearance The factory cars, as prototypes, will use racing rubber, which has a smaller rolling radius. But then they won't be measured as they'll race as prototypes.

This latest Porsche for real racers differs in other ways from the Abarth Carreras, which it nominally replaces. The 'mid-ships engine mounting is only a start. Their own engineer named Schroeder not only provided a whole new body style, he came up with a new/old chassis as well. It is new for Porsche, but the two-box-section beams down either side are not what you'd consider the last word in motordom. Still the car exhibited absolutely no lean, no tail-wagging and no slide at speed around the Solitude track near Stuttgart when presented to the Press and seemed very untwitchy for Linge to handle when we rode with him, in the rain at that.

Actually the track, which is public road on non-race days, carried normal two-way traffic, but they did have police out to give some measure of safety while the two test drivers ignored a speed zone at one corner. I doubt, however, if even the police were prepared for the abilities of this car. On my lap Linge passed a 45 mph limit sign turning an indicated 125 mph at the 7000 rpm red line in fourth gear, selected fifth went on up to 150 mph on the dial at 6500 rpm in top gear before getting on the brakes for the nasty corner the limit sign was supposed to be guarding. The policeman on corner guard was almost as pop-eyed as I.

On the other hand, the stocky Herr Linge trickled down to 1500 rpm in fourth gear and pulled away without so much as a gulp. I hardly think many of the GTS models will be purchased for the wife to take shopping, but they are about as tractable as any full racer I've ridden in. (Understandably, with traffic abroad. Porsche declined to let **any** journalists loose with 160 mph automobiles.)

There are yet more surprises in store for the GTS buyer with Porsche memories. For the first time fibreglass has become acceptable around Porsche, whatever other lands may have thought of it before. The substance is tradenamed Pantal. Heinkel, a name once known for other fast objects around Germany, does the actual moulding since Porsche has no facilities for such work, and the shells are then bonded directly to the steel chassis.

If you have a shunt with a 904, of any magnitude above the kind you can fix in fibreglass with some resin and a hot iron, you need a whole new body. As Porsche designer Tomala says, any race car which is really bashed should be replaced anyway, so they don't consider the unitary system a drawback on that score. More important, it increases overall stiffness by 50% over what you get with separate chassis and superstructure. The form itself owes a great deal to Porsche studies through the 1963

The chassis combines some of Porsche's oldest features with some exciting new ones. It weighs 330 lbs. less than the Abarth-Carrera.

160 MPH PORSCHE 904 GTS

season with notch-back GT cars. It uses the Kamm tail theory and a high-penetration nose.

Getting back to Linge, that worthy holds the Nurburg Ring class record of just under 10min. with a "normal" Porsche GT. He is reputed to have turned the Ring in 9.30 with a 904. Even allowing the usual discounts for factory watches, this is a fantastic time and should be giving buyers of the Abarth 2000, for instance, food for thought. A fabulous lightweight E Jaguar time, by comparison (faster than any official Ferrari GTO has done) is 9.38. The 2-litres are obviously on the way up.

A great deal of the gain in lap times, particularly on the 170-odd-curve Nurburg track, comes from an entirely reworked suspension system. In front, Porsche took over the coils and twin wishbones of their unhappily-discontinued GP eight-cylinder monoposto. In the back they didn't seem entirely sure themselves. While press notes referred to twin wishbones, all three cars shown on Solitude actually had a single, reversed wishbone on top in the rear, with a single lower tube. The brakes are outboard and the cast hub carriers set at an angle to the car's centre line. Twin radius arms lead forward on each side, again like their GP machine. A limited-slip differential is standard equipment.

Despite the husky-looking box frame and solid engine, the car weighs something like 330lb. less than the Abarth-Carrera model it should replace. With 10 litres of oil in the engine and another 3 litres in the gearbox, but no fuel in the 24-gallon tank, the 904 weighs right on 1435lb. This undoubtedly explains 0-100 acceleration times in not much over 11sec.

This sort of acceleration is aided no end by a typically solid Porsche gearbox, taken in the GTS case from their recent six-cylinder design. Naturally enough, the ratios are altered, along with the strength of the cogs and the casing, but the principle of the five-speed box remains. It is shifted by a rather longer than usual central lever—longer than most competition machines.

Which brings us to the driver's comfort and controls. The GTS—or 904, there seems to be some confusion around Porsche as to what they will call this one in the catalogues — is actually fitted out so you **could** drive it to a race. If your transporter broke down on the way, for instance.

The deep-dish bucket seats are fixed very firmly in place and non-adjustable. For a truly teutonic posterior they offer not only thigh support, but optimum lateral hold as well. For thinner drivers a bit of extra hip padding may be necessary to keep you from shifting in fast corners. The steering wheel and pedals furnish the adjustment range for various lengths of driver. The wheel may be moved up or back (but not changed as to rake) by a knob reached easily from the seat. The pedal bar requires a wrench and must be fixed for a given driver before starting off.

Instruments include a tachometer in the centre, easily seen through the wood-rimmed wheel and flanked by a speedometer and a combination dial for all other services. All are road-Porsche type. The car naturally carries full lighting equipment and even a good screen washer, while the wiper system consists of a single blade on twin, parallel arms. It clears a wide arc before the driver and even a portion of the passenger's window space. Incidentally, this is an honest GT car in seating as well, with equal provision for both passengers, unlike some "cheaters" in the class.

Noise is naturally a problem, with sound-proofing left to ear plugs for long distances. You **can** talk at speed if you have strong lungs. Furthermore, the engine and gearbox are reached comfortably by tilting the entire rear third of the bodywork back as a unit. For more major repairs, this can be disconnected in a very few minutes.

Porsche set out to provide a quasi-rally car with their vaunted durability produced in actual GT quantities, yet quick enough to win the 2-litre class in airport sprints. Even with the venerable four it does all of that—and they still have a six and an eight if the competition goes any quicker. They've come a ways since the VW days.

PORSCHE'S TARGA FLORIO

Absence of Works Ferraris and Failure of A.C. Cobras Results in an Easy Win for Porsche

Report and Photography by BILL GAVIN

SURPRISE VICTORY in the Targa Florio was achieved by a G.T. car—the works Porsche 904 *driven by Colin Davis, an Englishman resident in Italy, and Antonio Pucci, a local driver. Another Porsche* 904, *driven by Balzarini/Linge, finished second.*

COLIN DAVIS and Antonio Pucci won the 48th Targa Florio in Sicily last Sunday. Their car, and that of second placemen Balzarini and Linge, was a Porsche 904 GTS. Scuderia Ferrari absented themselves so that the score for supremacy in post-war Targas is now Porsche and Ferrari, five all. Victory in division 3 of the G.T. Championship went to the private Ferrari of Ferlaino and Taramazzo, which finished in fifth place and over 10 minutes ahead of the Dan Gurney/Jerry Grant A.C. Cobra-Ford, the only one of five Cobras to be classified. The homologated Alfa Romeo Giulia TZs finished third and fourth, so that the first prototype was the Barth/Maglioli Porsche 8-cylinder in sixth place, which had led after Bonnier's car had broken a halfshaft and until they broke a suspension damper.

THE Targa is without doubt the most beautiful circuit in the world, for Sicily in spring is delightful—wild flowers everywhere mixing with the new season's crops on the hill slopes, while the mountains which the 44.7-mile circuit encompasses are majestic in stature. The sun shone brightly throughout the several days of unofficial and one day of official practice and was at its best for race day. Unofficial practice consists of circulating as many times as possible in any car a driver can lay his hands on, from humble Fiat 600s to the 2-litre Porsche Carrera 2s imported from Stuttgart for the occasion, to the race cars themselves often employed by the lunatic fringe.

Among the hazards in unofficial practice are all sorts of beasts which stray or are driven across the road, truck drivers who use more than their fair share of the road or overtake in murderous situations, and the Sicilian peasants themselves, many of whom lack the intelligence to appreciate the danger of contact with several hundredweight of steel in motion. Poor Graham Hill hit a small boy who ran into the side of his practice car as it passed slowly through a village.

Official practice was held on Friday, and as usual chaos seemed to reign. Official timing, which can scarcely ever be faulted on race day, is never quite in gear for practice, and in those years when an official practice list is published it bears little relation to the truth. This year Bonnier was fastest in the open 2-litre prototype running its third Targa. His time for his one standing practice lap was 41 mins. 20 secs., which is very quick motoring. This car, which Jo was to share with Graham Hill, and the eight-cylinder 904 prototype of Barth and Maglioli were the most fancied runners for Maglioli (Targa winner 1953 and 1956) practised in 42 mins. 52 secs. Drivers usually only get one lap practice in the race car, for the 447 miles of uneven Sicilian roads which the cars must cover on the Sunday is considered a sufficiently gruelling test for most cars within a short space of time.

* * *

RACEDAY starts with everybody rising about 5 a.m., for most teams are based at Cefalu which lies some miles to the east of the circuit, and they must pass along part of the circuit to reach the pits before it closes at 6 a.m. The race itself started remarkably promptly at 8 a.m. when the first of the 1,150-1,300 c.c. class got flagged away. This class was as always dominated numerically by Alfa Romeos, ranging from Sprint Veloces to the ultra quick SZs (Speciale Zagato). But this year there were three Abarth-Simca 1300s, and from France, two René Bonnets and two Alpines.

The cars leave at 30-second intervals and at 8.09 a.m. the first of the 1,600 c.c. cars got away. One was an Alfa Giulia 1600, another an Osca, then the two works Alfa Romeo Giulias TZs shared by Bussinello/Todaro and Kim/Thiele, plus the private TZ of Nicodemi/Lessona. The TZs have been homologated since last November: they have an all-tubular chassis plus independent suspension all round, and a very clean-looking body.

Next cars away were the 2-litre G.T. models. Only non-Porsche cars were the works Abarth 2-litres of Herrmann/Patria and Spychiger/Riolo, and the private car of Conti/Venturi. There were no fewer than seven of the newly homologated Porsche 904s plus one of the 2-litre Le Mans coupés and a rather ropey Reutter bodied Carrera 2 driven by two Italians who hid their real identities under the mezzo-accurate pseudonyms of "Ben-Hur" and the "Tortoise". Pairings of the works 904s were Balzarini/Linge, Pucci/Colin Davis, and Klass/Neerpasch. Filipinetti entered Herbert Müller/Rayers, while Johnny Bulgari shared his car with Giovanni Grana. The Swedish 904 of Radeflak and Duneborn crashed in practice. Another private 904 was driven by Frenchmen Rey and Hanrioud, while the Italians Viannini and Deserti had the seventh 904.

Three Lancias were the only runners in the 2,500 c.c. class, all Flaminias, one bearing the standard Touring body, the others being Sprint Zagato models.

Seven Ferrari GTOs comprised the 3-litre class, the two re-bodied 1964 cars being driven by Ferlaino/Taramazzo and Facetti/Guichet. Standard GTOs were entered for Norinder/Troberg, Taormina/Tacci, Bourillot/"Bourbon Parma", Nicolosi/Zanardelli, and Ulisse/Fortinbras.

Likewise it was Cobras that filled the over 3-litre category. Phil Hill was paired with Bob Bondurant, Dan Gurney with Jerry Grant, Innes Ireland with Masten Gregory, while the third Shelby American car was lent to Alfa specialists Coco/Arena. Fifth Cobra was the privately entered car of Tommy Hitchcock III and Prince Zourab Tchkotoua.

A 2-litre class for Appendix C sports cars is always included in the Targa and usually includes some machinery that would nearly qualify as "cars of historical interest". This year we saw an old Maserati 2-litre, an elderly Osca, a Fiat V8, and a front-engined birdcage Maserati Tipo 61. Matched against this lot was a rear-engined V6 Ferrari for Terra/Toppetti and, for some odd reason,

NOT EXPECTED to do anything startling were the two rear-engined $2\frac{1}{2}$*-litre V8 A.T.S.s driven by Zeccoli/Gardi and Baghetti/Frescobaldi. They were last away at the start.*

AFTER AN IMPRESSIVE start, the A.C. Cobra threat gradually disappeared, the best the Anglo-American marque could achieve being a poor eighth. Innes Ireland's mount (above) was crashed by Masten Gregory. 1964 TYPE FERRARI GTO at speed (below). Minus a works team, Maranello were represented by private owners.

the 1,071 c.c. Mini-Cooper S of Harry Ratcliffe and Rupert Jones.

The main strength was in the 2-litre prototype category; the Bonnier/G. Hill machine was the 2-litre open car which first appeared here at the Targa in 1962. The flat-eight engine gives around 240 b.h.p. A similar engine was fitted to the Barth/Maglioli car which otherwise has mostly 904 specifications. Two Lancia Flavia prototypes with ugly bodies for Cella/Trotmann and Grosina/F. Frescobaldi were in this class. The little 1,300 c.c. Alpine-Renault was driven by Bianchi Bros., Mauro and Lucien. The Austin-Healey Sprite entered by Donald Healey for Paddy Hopkirk and Tommy Wisdom ran in this class, though it was a homologated model. Also incorrectly in this class was the standard M.G.B for Siddi/ Fondi.

Last away and least fancied were the A.T.S. prototypes of Zeccoli/Gardi and Baghetti/P. Frescobaldi. These cars have a tubular chassis with V8 2½-litre engine rear-mounted. Suspension is all independent with double wishbones at front and rear, the latter being augmented by a low radius arm. The Baghetti car departed at 8.33 a.m. precisely and it was 16 minutes before the first of the small cars appeared completing its 44.7-mile journey, during which it climbed to almost 2,000 ft. above sea level as the road snakes inland to the mountains and out again for the fast run along the coast.

Third of the small cars to pass was the Rosinski/Rolland Alpine which completed the first lap in the splendid time of 46 mins. 40 secs., over a minute faster than Laureti/ Ridolfi (Abarth-Simca), who went on to win the class after Rosinski did a 45 mins. 28 secs. lap and then lost a wheel and damaged the suspension. Deserti did the fastest standing lap of the 904s, and his time of 42 mins. 49.3 secs. had him placed sixth overall at this stage, but the car lasted only two more laps. Guichet put in an even faster opening lap at 42 mins. 34.1 secs. in the GTO to lie fourth overall, after shattering by something like a minute the best ever GTO time, and on the standing lap at that! Gurney opened with a 42 mins. 23 secs. and Phil Hill with a 42 mins. 35 secs. which placed them third and fifth. Bonnier's 41 mins. 16.1 secs. was an excellent opening lap—unfortunately it was his last, for on the second lap a halfshaft U.J. broke, letting Barth through into the lead.

Barth's lead lasted only three laps, during which time he had built up a lead of 2 mins. 40 secs. on Bulgari in the first of the 904s. Guichet was really pushing along the GTO and had lapped in 41 mins. 13 secs. to occupy second place at the end of lap two. Transmission trouble later slowed the car, causing its retirement after five laps. On lap five Maglioli (who took over from Barth after three laps) limped into the pits, a pin retaining the shock absorbing having sheared. It took over 20 minutes to sort this out, and by this time the car was out of the running. Bulgari had taken over the lead, but at the end of the sixth lap he came into the pits, his brilliant drive over, for the rear suspension mountings of the 904 had pulled out of the chassis. This let the Pucci-Davis car through into the lead, the Englishman now at the wheel.

The Gurney Cobra, after lying third on laps two and three, dropped back as it became difficult to obtain second and top gears on the four-speed Ford unit. Dan had lapped the Cobra in 42 mins. 13 secs., but the cars were difficult to drive, the hard suspension really bouncing the cars about the road on this bumpy circuit. But by the end of the sixth lap Grant had the Cobra back in second place, and with the Hill/ Bondurant car in fourth place behind the Klass/Neerpasch Porsche it seemed possible that the Cobras might pull off the Targa at their first attempt. Fifth was the Arena/Coco Cobra, while the other Shelby car of Ireland/ Gregory and the Tchkotoua/Hitchcock car both had long stops to repair damaged suspensions.

The British entries weren't faring so well—Hopkirk got the little 1,100 c.c. Sprite around in less than 45 minutes before the axle broke, while the Mini-Cooper S shed a wheel and damaged the suspension.

By the end of the eighth lap Pucci and Davis had a seven-minute lead over team mates Klass and Neerpasch. Gurney was still nobly pushing the Cobra along 10 minutes behind the leader, but half a lap later the suspension broke. But still he pressed on, with the car handling badly. The Bussinello/ Todaro Alfa Romeo TZ had moved right up to fourth place, while a minute and half behind them came the Linge/Balzarini Porsche 904, advancing through the field after a lengthy stop at the end of lap 1.

Sixth was the "Kim"/Thiele Alfa, a minute ahead of the leading GTO, that of Ferlaino and Taramazzo. Barth and Maglioli had moved through to eight place ahead of the Bourillot/"Bourbon Parma" and Norinder/ Troberg GTOs.

As the race ran out it seemed obvious that Pucci could coast in—the Klass/Neerpasch car ran out of brakes while Linge brought another 904 up to second place. The Gurney/ Grant Cobra dropped to eighth place with its faulty suspension, though similar trouble had eliminated both the Hill/Bondurant and Coco/ Arena car. This let the little Alfas through into third and fourth place, with the Ferlaino Ferrari taking fifth spot a minute and a half ahead of the Barth/Maglioli prototype.

The race was a disappointing one; first because of the withdrawal of Ferrari and secondly because of the almost total failure of the Cobras.

LITTLE Mini-Cooper S of Harry Ratcliffe/Rupert Jones lost a wheel and suffered suspension damage sufficient to put it out of the race.

RESULTS

1, Antonio Pucci/Colin Davis (Porsche 904), 7 h. 10 m. 53.3 s., 62.30 m.p.h.

2, Gianni Balzarini/Herbert Linge (Porsche 904), 7 h. 23 m. 15.3 s.

3, Roberto Bussinello/Nino Todaro (Alfa Romeo Giulia TZ), 7 h. 27 m. 7.0 s.

4, "Kim"/Alfonso Thiele (Alfa Romeo Giulia TZ), 7 h. 27 m. 38.2 s.

5, Claudio Ferlaino/Luigi Taramazzo (Ferrari GTO), 7 h. 28 m. 25.0 s.

6, Edgar Barth/Umberto Maglioli (Porsche "8"), 7 h. 29 m. 16.4 s.

7, Gunther Klass/Jochen Neerpasch (Porsche 904), 7 h. 30 m. 45.2 s.

8, Dan Gurney/Jerry Grant (A.C. Cobra-Ford), 7 h. 38 m. 5.0 s.

9, Ulf Norinder/Picko Troberg (Ferrari GTO), 7 h. 40 m. 44.0 s.

10, Claude Bourillot/"Bourbon Parma" (Ferrari GTO), 7 h. 46 m. 45.0 s.

PORSCHE 904 GTS

The chassis is new, the glass-fiber body is new, the suspension is new... only the two-liter Carrera engine is familiar, and it now purrs out 200 SAE horsepower!

ROAD TEST 6/64 SPORTS CAR GRAPHIC

PORSCHE HAS OFTEN BEEN ACCUSED, LATELY, OF BECOMING TOO CONSERVATIVE. Among other things, it is true that until the presentation at Frankfurt of their new two-liter six-cylinder, their line of cars has remained basically unchanged for the past ten years. It is also true that Porsche, whose name is based on competition, gave up Formula I racing on the eve of success because of what I would call a certain lack of enthusiasm or fighting spirit. Fortunately in November 1963 Porsche came up with their new Type 904 GTS, a car showing strikingly that this famous though small firm had lost neither its touch nor its imagination for cars in the high performance field. Indeed, with this new competition GT car, Porsche has come out with an exciting design, very up to date in conception, and certainly different in many regards from the GTs they have built in the past. Chassis, body and suspension are all new; a modified version of the well known two-liter four-cylinder Carrera engine has been kept as the power plant for this otherwise all new 904 GTS.

Apparently, rather than using the elaborate and expensive eight-cylinder engine (used for both Formula I and sports car racing) to power this new car, the Porsche engineers decided to stick to their faithful four-cylinder, four overhead cam engine, especially since a new design of the head, different type pistons and a new camshaft were able to raise the 160 horsepower of the racing GT version of this engine to a very satisfying 180 hp DIN (200 SAE). The maximum torque is raised from 125 lb/ft at 4,800 rpm to 148 at 5,000 rpm. With this competitive 180 hp obtained, it was necessary to build around it a lighter car with a more up to date suspension and a new streamlined body, capable of giving those precious extra miles per hour in top speed, as well as cutting down the drag factor in a more conventional GT type body. Instead of making a tubular type chassis for the 904 (very good but expensive to build, even produced in a certain quantity), Porsche made a simple but very sturdy box section type frame.

The typical Porsche suspension — torsion bars, trailing arms and swing axle on the rear — was abandoned for the 904 in favor of an all around independent suspension directly derived from their Grand Prix cars. In

Leather seats are contoured and molded into the body.

A powerful, single wiper sweeps the curving windshield.

Production is well under way, with homologation due by LeMans.

PORSCHE 904 GTS

Lucky Bernard beat us to another fine road test. Rats!

order to save weight and money a glass fiber body was selected for the first time in Porsche history. This body, which was designed by the eldest son of Mr. Porsche, is composed of two main sections. The front one includes, in one moulded unit, the front end of the car, the cockpit and the top, while the rear unit includes the whole rear end of the car, hood and fenders. The front section is fixed while the rear is quickly removable giving full access to the engine and rear suspension. Made of strong glass fiber, the body itself looks very rigid and, with the main section being of one moulded piece, you can say that the 904 uses a monocoque type body fixed to a frame. The lines of this body are quite exciting, and knowing the importance which Porsche gives to streamlining, you can be sure that past racing experiences have been used to good advantage, as well as careful wind tunnel studies.

On the scales the 904 shows an interesting weight of 1430 pounds, with 11 quarts of oil and a lot of air. This is about 250 pounds lighter than their 1963 GT competition car, and 600 less than the production Carrera G-S. Considering this weight saving, plus extra horsepower, plus the extremely good shape of the car, it is not difficult to assume that the new 904 is a much faster car than the not-so-slow 1963 GT model. This assumption was quickly clarified for us when Porsche Racing Manager Von Hanstein arranged an exclusive test of

PORSCHE 904 GTS

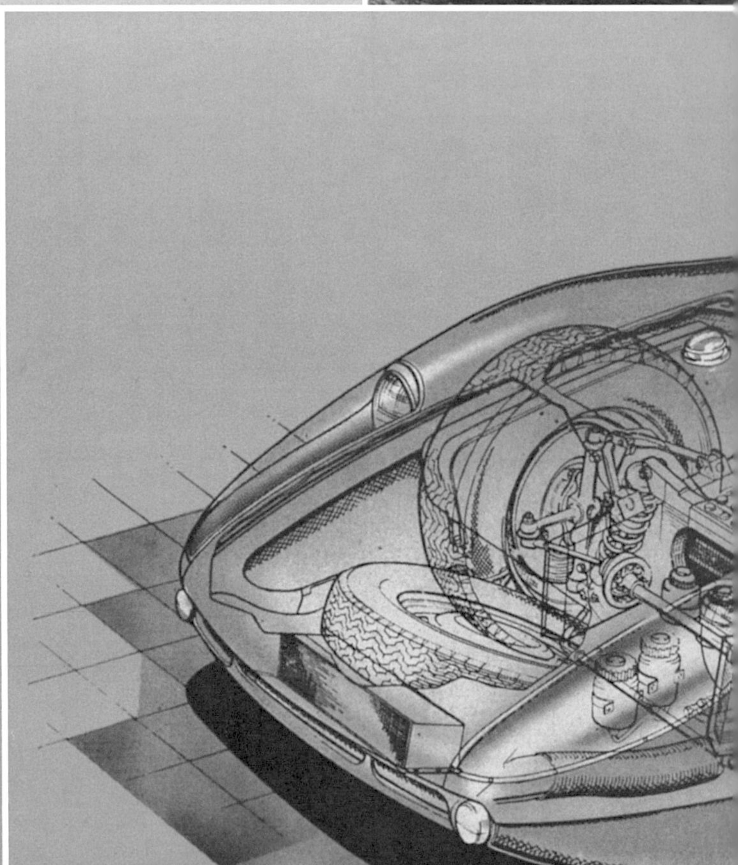

Porsche departs from traditional designs in many respects as is evidenced by tall, box-section frame, unusual suspension layout, and swivel inner U-joint. Glass fibre body is also something radical for the firm. Tires are Dunlop SP's.

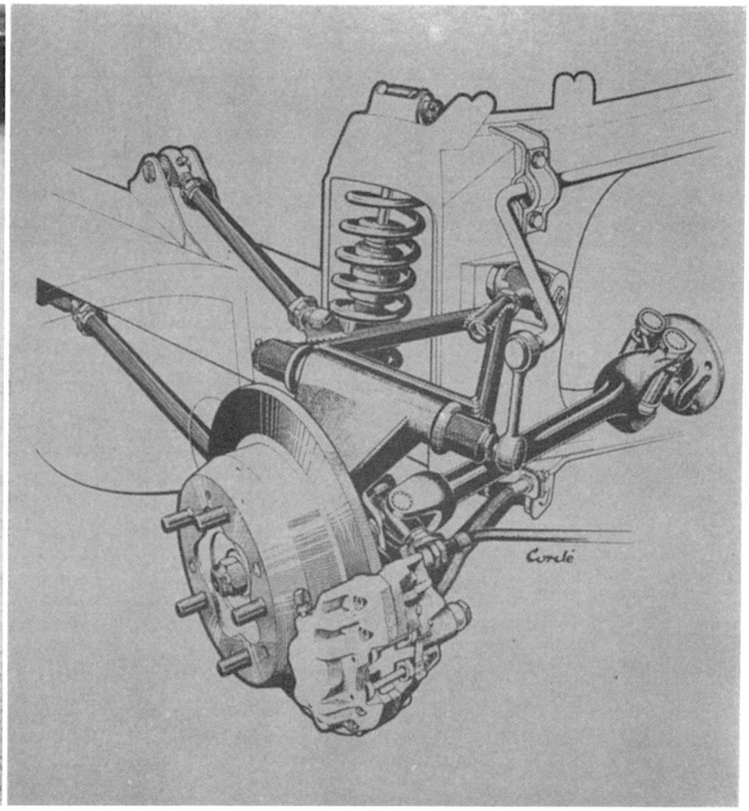

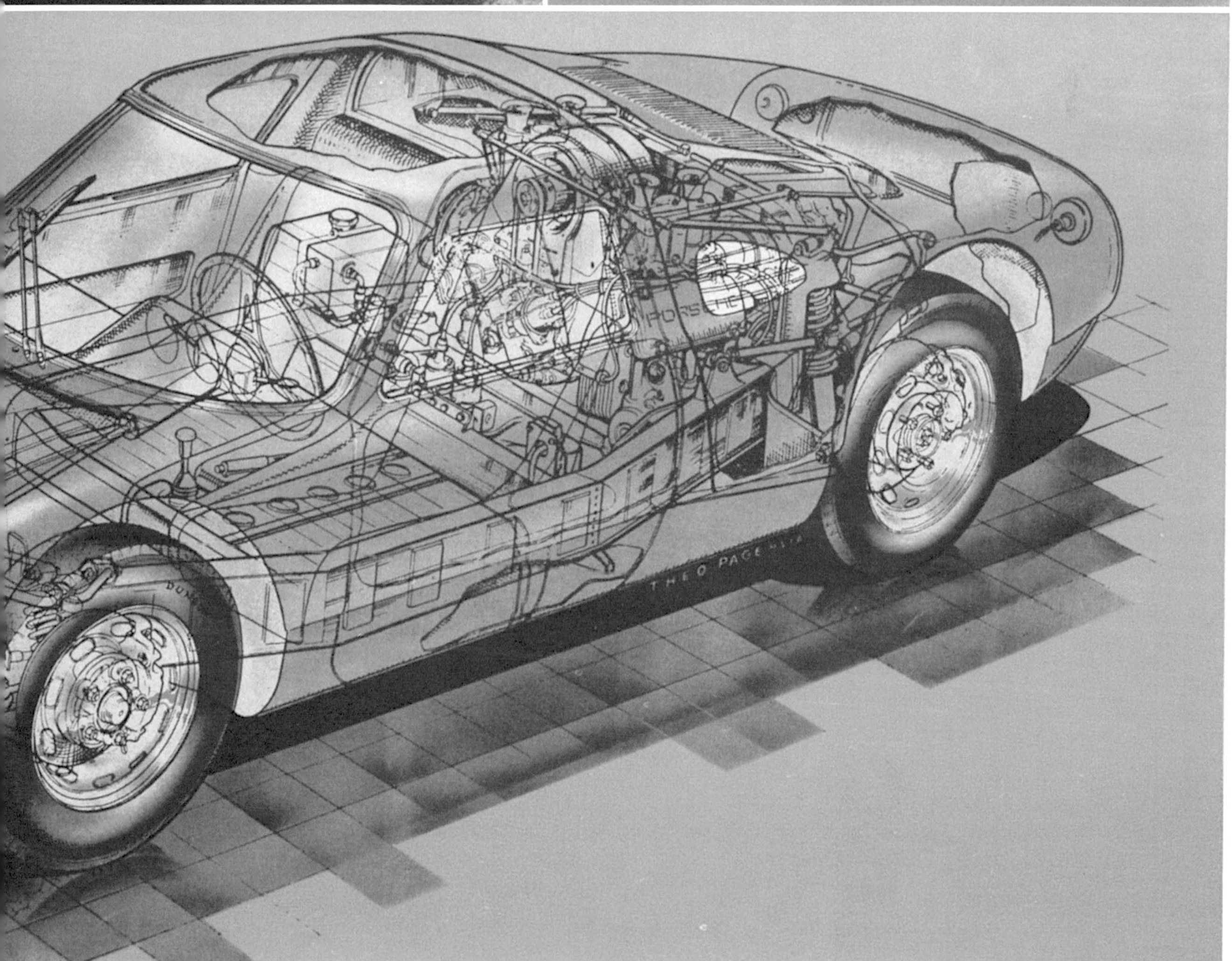
PORSCHE
THEO PAGE

PORSCHE 904 GTS ROAD TEST 6/64

Vehicle .... Porsche
PRICE (as tested) .... Approx. $7400 @ factory
MODEL .... 904 GTS
OPTIONS .... Unknown

ENGINE:

Type .... Four-cylinder, air-cooled (dry sump)
Head .... Four over-head cams
Valves .... Direct-acting
Max. bhp .... 180(DIN) @ 7,000 rpm
Max. Torque .... 148 lbs. ft. @ 5,000 rpm
Bore .... 3.62 in. 92 mm.
Stroke .... 2.91 in. 74 mm.
Displacement .... 120 cu. in. 1966 cc.
Compression Ratio .... 9.8 to 1
Induction System .... Two double-throat Weber downdrafts
Exhaust System .... Four-into-one headers, racing version; twin exhaust for street version
Electrical System .... 12 v. twin ignition distributor

CLUTCH:

Dry, single disc
Diameter .... 7.8 in.
Actuation .... Mechanical

DIFFERENTIAL:

Ratio .... 4.43 to 1
Drive Axles (type) ... Independent, open

TRANSMISSION:

Five speed forward, all synchronized
Ratios: 1st .... 2.643 to 1
2nd .... 1.684 to 1
3rd .... 1.318 to 1
4th .... 1.040 to 1
5th .... 0.821 to 1

STEERING:

Rack and pinion
Turns Lock to Lock 2.25
Turn Circle .... 38 ft.

BRAKES:

Drum or Disc Diameter .... N.A. in.
Swept Area .... N.A. sq. in.

CHASSIS:

Frame .... Box section, mono-beam, steel
Body .... Glass fiber in two main pieces
Front Suspension ... Unequal arms, anti-sway bar, tube shocks, coil springs
Rear Suspension ..... Independent unequal wishbones, twin trailing arms, tube shocks, coil springs
Tire Size & Type .... Dunlop SP and Racing 600 x 15 and 550 x 15

WEIGHTS AND MEASURES:

Wheelbase .... 90.5 in.
Front Track .... 51.5 in.
Rear Track .... 51.4 in.
Overall Height .... 41.6 in.
Overall Width .... 60.5 in.
Overall Length .... 157 in.
Ground Clearance .... 2.8 in.
Curb Weight .... 1430 lbs.
Test Weight .... 1680 lbs.
Crankcase .... 11 qts.
Cooling System .... air-cooled
Gas Tank .... 29 gals.

PERFORMANCE:

0-30 .... 2.6 sec.
0-40 .... 3.4 sec.
0-50 .... 4.5 sec.
0-60 .... 5.6 sec.
0-70 .... 6.5 sec.
0-80 .... 8.4 sec.
0-90 .... 10.2 sec.
0-100 .... 12.4 sec.
Standing ¼ mile .... 13.7 sec. @ 103 mph
Top Speed (av. two-way run) .... 158.5 mph

Speed Error	30	40	50	60	70	80	90
Actual	30	40	50	60	70	80	90

Fuel Consumption:
Test .... N.A. mpg
Average .... 11 mpg

Recommended Shift Points at 7,000 rpm:
Max. 1st .... 47 mph
Max. 2nd .... 74 mph
Max. 3rd .... 102 mph
Max. 4th .... 130 mph

RPM Red-line .... 7,500 rpm
Speed Ranges in gears:
1st .... 0 to 47 mph
2nd .... 6 to 74 mph
3rd .... 13 to 102 mph
4th .... 20 to 130 mph
5th .... 35 to 158.5 mph

Brake Test: No fade encountered on several stops.

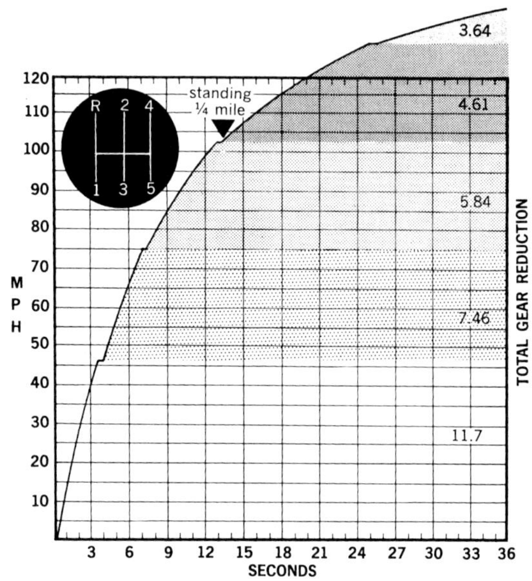

REFERENCE FACTORS:

Bhp per Cubic Inch 1.5
Lbs. per bhp 7.22
Piston Speed @ Peak rpm .... 3395 ft./min.
Sq. In. Swept Brake area per Lb. .... N.A.

this exciting newcomer for SCG. This test was made around Stuttgart on the open road, on the Solitude race course and, of course, on the autobahn. During our two days stay the weather was not especially pleasant, with rain and snow too often interfering, but it proved to me that the 904 is a fantastic car in the wet, that you can get quite damp in it, and that an unwanted inside breeze can keep your feet very cold in the Wintertime. Mind you, there was a heater, a luxury for a competition car, but it seemed to work mostly around the upper part of my body. But those are minor comfort details and what was most important was to see that this car could be driven at very high speeds (I was usually cruising around 130 mph on the autobahn) on wet roads in a perfectly relaxed manner. This was due of course to the remarkable general handling qualities of the car, to its fine steering and excellently balanced, powerful disc brakes, but it was also due, during this wet test, to the best windshield wipers I have ever seen on any car. Windshield wipers seem to be one of the things manufacturers too often neglect, and in the immense majority of today's cars the wipers are terribly inefficient at speeds over 70 or 80 mph. Sadly enough this point applies also to GT and sports cars, usually driven fast by their owners. In the case of the 904 there is only one wiper, but it is such a powerful one (bus type) that at 130 mph on the autobahn I had no vision problems and this was true even when being really splashed by cars I was overtaking. This reminds me of another quality of the car in the rain at high speeds. The shape of the 904 (it is very low, with a total height of 41 inches) is such that most of the water thrown back at it by other moving cars ahead does not seem to hit the windshield, but goes over it. I remember following a Mercedes closely at a time. It was doing a good 90 mph and was passing another car which meant that I had to stay behind for a certain period. Well, all the water thrown back by both of them did not obstruct my vision at all, as it normally would. What a wonderful change, and what a feeling of safety it gave. One criticism — I believe it would be better to have two wipers which would give better corner vision when driving on curvy roads.

When I took over the 904 GTS in the Porsche competition department, it was the first time I had actually seen the car, as I was away at the time of its official presentation in late November. The photos showed an unusual, fast looking machine while in reality it is even more exciting. Only one real disappointment, the finish of the inside of the car is not up to the standards of, for example, the two-liter Abarth GT. I take into consideration the fact that the car I drove was one of the very first prototypes, but production models, although much

PORSCHE 904

better, are still not as well finished as they should be. Granted it is a competition car, but some of them are actually being sold for street use, with a milder engine and quieter exhaust, and so why not make the interior more attractive with a neater general finish.

Getting into a 41-inch-high coupe imposes some problems to girls wearing tight skirts, but any supple sportsman (ha!) will not find the operation too painful. Anyway, after a little exercise you are rewarded as the sitting-driving position is very good and the cockpit roomy enough to accommodate six footers comfortably. The seats are not adjustable as they are part of the molded front unit of the car's body. Therefore a suitable driving position is found by adjusting the three pedals via a sort of short hand brake type lever which changes the length of the pedals. The steering is also adjustable and so there is no problem in finding a near ideal driving position, especially with the well designed, nicely padded seats. I said above that the cockpit is very simple, but the black dashboard shows a complete line of instruments and the various gadgets are easy to reach. As expected the short stick shift commanding the all-synchronized five-speed gear box is well placed and extremely easy to use. The gear pattern is different, but not enough to be any sort of problem. Actually I only goofed once when I started in third gear instead of first. I was leaving the garage and the mechanics were looking at me with a certain irony. You know, a journalist ... Anyway, neither the clutch nor the engine seemed to be in too much pain, and I went on through the gate in the wrong gear, which proved that the 904 can be driven off in third gear, proving also the flexibility of this highly tuned engine.

Driving the 904 does not present any special problems, and this simply because it is a very "clean cut" car, as well as a very forgiving one. The clutch, which requires a certain pressure, is very direct, but with enough progression that you can feel it. This quality, associated with an engine which is civilized (for a competition car) allows you to make clean starts (I had by then sorted out the first gear) without either stalling or overdoing things. With the two-liter Abarth, for example again, quick, clean starts were difficult as you either used too much or not enough throttle. It is no problem with the 904 either in town or during acceleration runs. This brings to light the fact that the 904 can actually be driven in traffic, which is why a street version is being offered to those who want to be at the wheel of something different. Different it certainly is, in looks and in feeling, and everywhere I went crowds immediately gathered around the car, most of the people looking quite amazed and wondering what it really was. "*Das ist ein Porsche*" was my simple answer to questioners who then seemed satisfied, and certainly much less surprised.

When doing high speed tests, during which the car was eventually clocked at a very honorable 158 mph, the 904 remained extremely stable with no wandering feeling, even when at maximum speed. Driving in the rain is never amusing and I had it most of the time spent with this car. Well, as I said earlier, I never had any uncomfortable experiences with the 904, and I think that this is self explanatory of the basic qualities of this car. Emergency stops on the wet at speeds of 125 mph were gratifying, as the car not only slowed down very quickly but it always did it without any trace of swerving, which often happens with disc brakes if you drive in the rain for a long time without using them. Those discs on the 904 seemed perfect, the only complaint being that you have to push quite hard on the pedal for them to function efficiently. They are really a man's brakes. I did not encounter fading on my test but my friends Uli Wieselman and Paul Frere, who drove the same car earlier, had some fading showing at a time on some fast, winding roads. Uli Wieselman, incidentally, was very kind in completing some of the acceleration figures which I could not take because of the weather. Amusingly, our figures were generally a little better than those recorded by the factory. Perhaps our car had loosened up by the time we tried it. These figures are very impressive and place this two-liter car (only 120 cubic inches) not only in the top of its class, along with the two-liter Abarth (interestingly enough the two are extremely close to each other, the Porsche having slightly better acceleration but the Abarth being about three miles an hour faster) but place it extremely well in the bigger GT classes as well. For example, it is quicker in acceleration than the three-liter GTO Ferrari which won the GT

Championship last year. The 904 will accelerate from zero to 50 mph in 4.5 seconds, zero to 60 mph in 5.6 seconds, to 80 mph in 8.4 seconds, and to 100 mph in 12.4 seconds, while the standing quarter of a mile and standing kilometer were done in, respectively, 13.7 seconds and 24.5 seconds. Really not bad at all for a two-liter car!

In the car we drove the gear ratios were excellent, with 1st going to 47 mph, 2nd to 75, 3rd to 102, and 4th to 130. Shifts were fast and easy to do, and that did not surprise me, knowing the fine reputation that Porsche has in the gearbox field. With such gear ratios you are well placed on just about any type of road. I drove on quite a number of various types, mostly with good surfacing. The 904 was a joy to drive, with its light controls, fine brakes and superb handling. The car seemed to be just as much at ease on tight or fast turns, and its basically neutral behavior makes it very easy to drive, especially since some slight oversteering appears as a warning when pushing it harder. On the rapid Solitude course, which was very slippery that day, I was amazed to see how fast you could take it around corners. The car really felt as if it were glued to the road and with it you could take a very clean line which the car followed faithfully. Sometimes on very slippery patches you could feel the back going, but a quick correction of the very direct steering, and a gentle use of the throttle, brought it back under full control in a flash. Generally speaking, the 904 truly feels as if it is on rails, very much like a 1963 single seater Formula car, except that it is bigger. Considering the type of car it is, the ride was very acceptable. The inside noise, however, was on the high side, which can be explained by the fact that the engine is directly against the cockpit (it is mounted ahead of the back axle) and that the exhaust is a very large, noisy one. I know some of the corners of the Solitude course quite well, having tested cars there before. I remembered at what speeds I have taken these corners with fast cars and thus it took me some time to realize that I was going just as fast, but that I was driving on a wet track. That 904 is really quite a sensational car, and those SP Dunlops are terrific in the wet.

As far as I can see, Porsche has made a major step forward with their 904 GTS and their efforts should be rewarded by many successes this year. Priced at about $7400 at the factory, the 904 should also be a great commercial success. I would not be a bit surprised if several hundred of them are sold, which is quite a number for such a highly specialized type of car which can hardly be used for normal touring simply because there is no space for baggage. The 904 GTS is a well born machine. It is exciting looking, different, stimulating to drive, it will give many happy moments to its owners on the road and on the tracks, as well as giving refreshing new prestige to Porsche. – *Bernard Cahier*

PORSCHE 904

Porsche continues to offer the private owner an outstanding value in the competition field

PHOTOS BY GORDON CHITTENDEN

WHEN THE MODEL 550 Porsche appeared in American road racing in the mid-Fifties, it brought a new dimension to the sport in this country. Prior to its becoming available to private owners, campaigning a 1500-cc sports/racing car was almost certain to lead to frustrations and anxieties. In those days there weren't many 1.5-liter racing cars that could be called "factory built" and even these few tended to be frightfully expensive, yet still had only the spottiest factory support for the private owner. Not only was there no assurance that a completely different model would not obsolete yours at any moment but there was also a good chance you couldn't obtain the proper replacement parts to keep it running anyway. As a result, the most successful 1500-cc cars were locally produced specials that coupled a tuned production car engine with a lightweight chassis and body.

The Porsche 550 changed all that with the introduction of a 110-bhp, dohc 1498-cc engine at the Paris Show in the autumn of 1953 which heralded the revolution. Within a year, the MG Specials were virtually out of contention and, with a few notable exceptions, the Porsche 550 and its derivatives completely dominated the under-2-liter sports/racing class right up until the arrival of the twin-cam 1600 Lotus 23-B in 1963.

In fact, the Porsche 550 became a sort of Everyman's Racing Car, the first Everyman's Racing Car in American racing. For a minimal amount of money, less than $8000 (which Everyracingman considers cheap), it was possible to purchase a soundly engineered, thoroughly developed and responsibly supported racing car. You could buy one, race it with a good chance of success, depending on your own driving skill, be assured that a new model wasn't going to make yours old-fashioned next week, and you could get the spare parts you needed. The 550 even brought a word into popular use in American racing—Spyder. Ah yes, Spyder. What a nice foreign sound it made in the mouth, Spyder with a "y."

But things have changed since 1954. In those years, the

ROAD TEST

PORSCHE 904

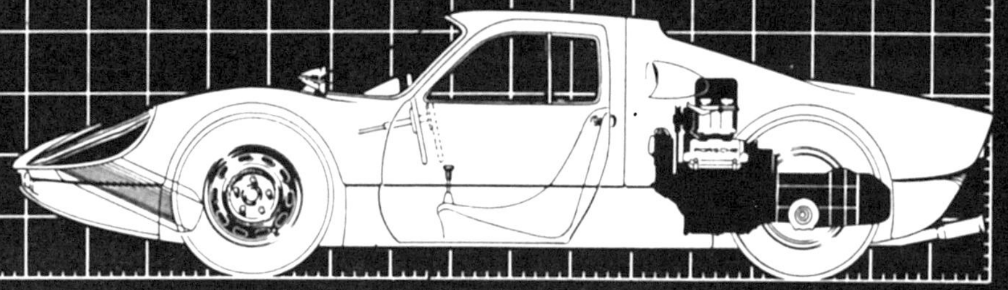

PRICE

List, FOB Stuttgart........$7425

ENGINE

Engine, no. cyl, type: Air-cooled flat 4 dohc
Bore x stroke, in......3.62 x 2.91
Displacement, cc............1966
Equivalent cu in............120
Compression ratio...........9.8:1
Bhp @ rpm..........198 @ 7200
Equivalent mph..............150
Torque @ rpm, lb-ft.144.5 @ 5000
Equivalent mph..............97
Carburetor, no. make...2 Webers
No. barrels, dia.......2-46 mm
Type fuel required......Premium

DRIVE TRAIN

Clutch diameter & type:...7.9 in., Single plate dry.
Gear ratios, 5th (0.889)......3.936
4th (1.040)...............4.605
3rd (1.217)...............5.389
2nd (1.550)..............6.863
1st (2.643)..............11.703
Synchromesh.............on all 5
Differential, type & ratio: Spiral bevel, 4.428:1
Optional gearbox ratios: Ratios shown are those used at Sebring. Four other "standard" sets available to give final drive ratios of 4.605, 4.260, 3.636 and 3.362:1.

CHASSIS & SUSPENSION

Frame type: Box section ladder frame.
Brake type..................Disc
Total pad area, sq in......28.68
Tire size, f/r.......550-15/600-15
Wheel revs/mi...............785
Steering type.......Rack & pinion
Overall ratio..............15:1
Turns, lock to lock.........2.07
Turning circle, ft............42
Front suspension: Independent with double A-arms, coil springs, tube shocks, anti-roll bar.
Rear suspension: Independent with reversed A-arms, trailing links, coil springs, tube shocks, anti-roll bar.

ACCOMMODATION

Normal capacity, persons........2
Hip room in..............2x18.5
Head room...................38
Seat back adjustment, deg......0
Entrance height, in...........41.5
Step-over height.............15.0
Floor height..................5.5
Door width..................30.0
Driver comfort rating:
for driver 69-in. tall.........80
for driver 72-in. tall.........60
for driver 75-in. tall.........45

GENERAL

Curb weight, lb (no fuel).....1350
Test weight.................1825
Weight distribution without driver, percent..........41/59
Wheelbase, in..............90.55
Track, front/rear......51.73/51.65
Overall length.............161.02
Width.....................60.62
Height....................41.93
Frontal area, sq ft.........14.00
Ground clearance, in.........4.72
Overhang, front............39.37
Rear......................31.10
Departure angle, no load, deg...14
Usable trunk space, cu ft......7.5
Fuel tank capacity, gal.........29

INSTRUMENTATION

Instruments: 280-kph speedometer, 8000-rpm tach, oil pressure, oil temperature, fuel.
Warning lamps: Turn indicators, generator, high beam, parking light, oil pressure, fuel level.

MISCELLANEOUS

Body styles available: Coupe as tested.

CALCULATED DATA

Lb/hp (test wt)...............9.2
Cu ft/ton mi (test wt)........119
Mph/1000 rpm (4th).........19.4
Engine revs/mi.............3090
Piston travel, ft/mi..........1499
Rpm @ 2500 ft/min.........5154
Equivalent mph.............100
R&T wear index.............46.4

MAINTENANCE

Crankcase capacity, qt........10.6
Change interval, mi........3000
Oil filter type...........Full-flow
Clean interval, mi.........3000
Lube interval, mi............3000
Tire pressures, front/rear, psi.....................32/35

ROAD TEST RESULTS

ACCELERATION

0-30 mph, sec................3.2
0-40 mph...................4.0
0-50 mph...................5.2
0-60 mph...................6.4
0-70 mph...................7.8
0-80 mph...................9.6
0-100 mph.................13.7
0-120 mph.................20.0
Passing test, 50-70 mph.......2.5
Standing ¼ mi, sec..........14.5
Speed at end, mph.........103

GRADE CLIMBING

(Tapley Data)

Data not taken

TOP SPEEDS

5th (7200), mph..........est 150
4th (7200)..................120
3rd (7200)..................101
2nd (7200)..................80
1st (7200)..................47

FUEL CONSUMPTION

Normal range, mpg...........na
Cruising range, mi............na

SPEEDOMETER ERROR

Speedometer calibrated in kph

Data not taken

ACCELERATION & COASTING

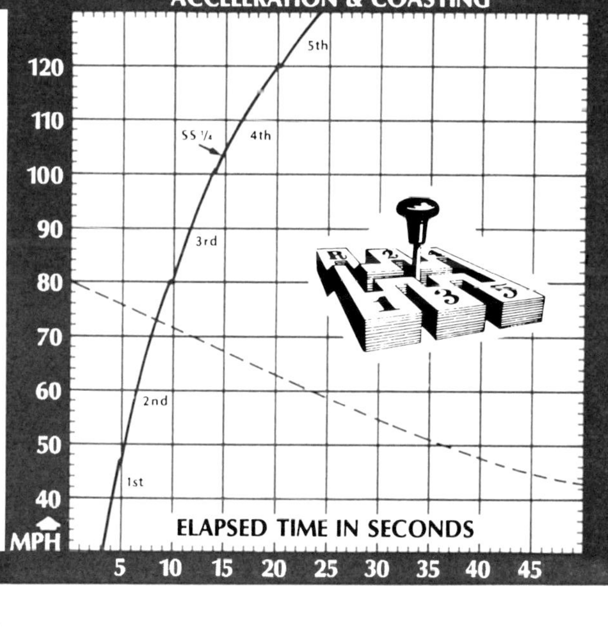

Fifteen-inch wheels were retained by Porsche for the 904 so that it would be a practical road car suitable for rallying in all but the worst territory

Porsche 904

QUITE the most interesting Porsche to leave the Zuffenhausen factory in the suburbs of Stuttgart for some time is the Type 904 Carrera GTS which was announced towards the end of last year. In a number of significant respects the design breaks new ground for Porsche. The most interesting departure perhaps is the introduction of a separate chassis frame—Porsche cars have, of course, hitherto employed unitary construction. The body, furthermore, is a glass-fibre, reinforced-plastic structure of the unusual shape shown in the illustrations.

Engine and transmission are typical closed Porsche units, but for the first time in a car available to the normal customer, the engine is immediately behind the occupants and forward of the final drive. It is also the first competition car from Zuffenhausen to have disc brakes all round, these being of the Porsche-ATE dual circuit type. They are mounted outboard and are manufactured in Germany under Girling licence, neatly incorporating a small drum brake at the rear for parking purposes. This type of handbrake, it may be remembered, was introduced some time ago by Porsche when they first offered discs.

The frame is a light and very stiff box section ladder-type structure with light but stiff bulkheads fore and aft and two transverse box-section cross members located as shown.

The dimensions of the 904 show it to be a compact vehicle. To provide maximum space for the two occupants the frame members are bowed out between the wheels; yet further space is ingeniously contrived by slightly chamfering the frame members. The frame structure is already very rigid, but additional stiffness is achieved by bolting the non-ferrous body to the frame members and permanently bonding the units together with adhesive.

In the interests of lightness the two-piece sliding windows are of perspex and the wide and deeply raked windscreen is provided with an unusual single-blade wiper with twin arms and mechanism to provide a wide parallel sweep across the screen. Our illustrations show the suspension in which it can be seen that the unequal length wishbones with co-axial coil-spring damper units are similar to the layout used by Porsche on their grand prix car. Equal length transverse links and unequal length trailing arms are employed at the rear, the

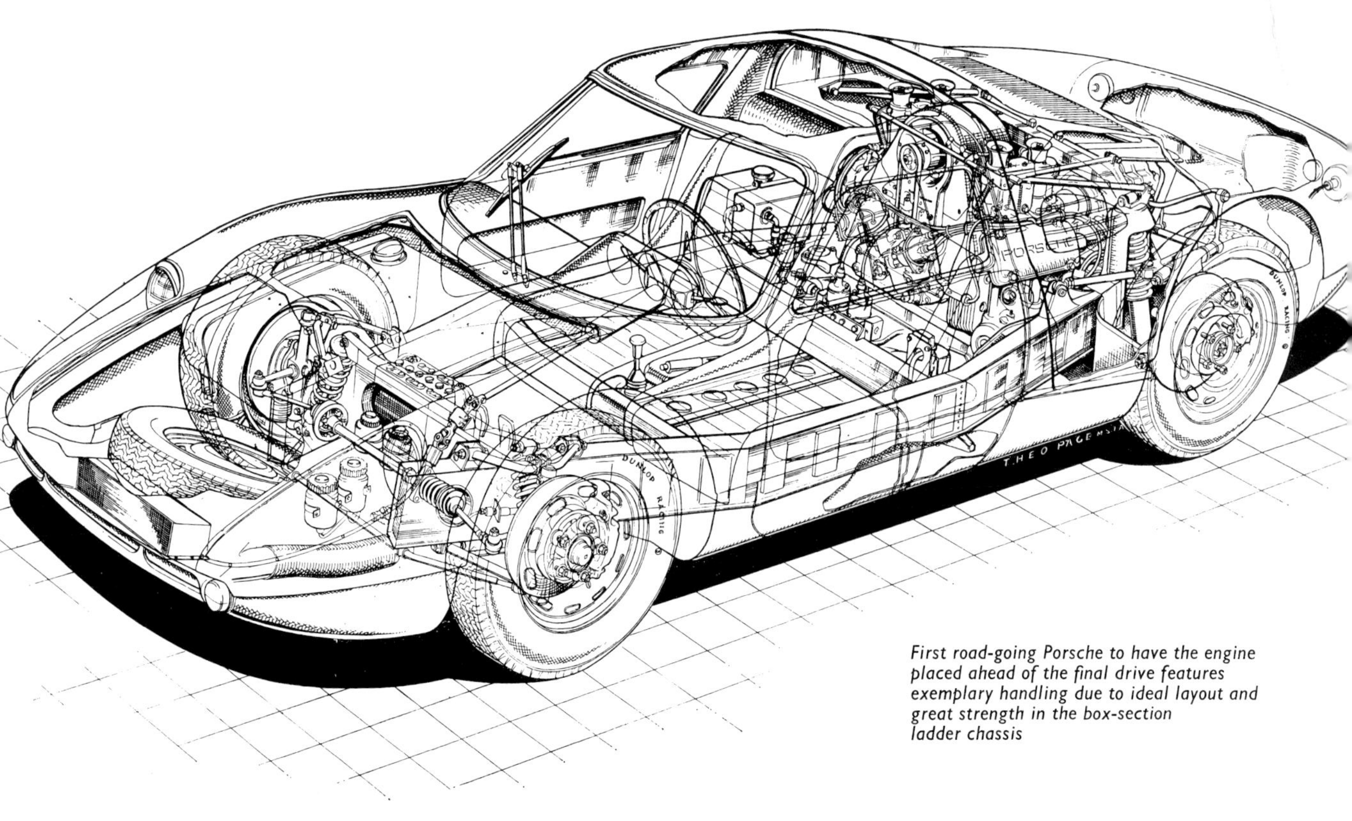

First road-going Porsche to have the engine placed ahead of the final drive features exemplary handling due to ideal layout and great strength in the box-section ladder chassis

Side members of the chassis are bowed outwards to provide maximum space for two people inside the car; low mounting of the power unit is apparent

arrangement providing a small amount of required rear wheel steering.

Turning to the engine, all the 904 cars so far built have been provided with the four-cylinder, two-litre, twin-overhead camshaft engines, but the designers have allowed for either their six-cylinder or eight-cylinder engines to be incorporated in the frame without difficulty should it be decided to race these units as factory entries. So far as we are aware, however, there is no thought of selling the car with these engines and to the best of our knowledge the 155 bhp at 6,400 rpm standard-rated engine has not so far been supplied; all the cars sold at the time this was written had, we believe, been provided with the racing version which provides 180 bhp net at 7,000 rpm.

Although the engine is based on the Carrera twin-cam power unit, the alloy engine has been considerably developed in order to provide the power and torque necessary for the new car. There are a pair of 46 mm twin-choke Weber carburettors, one for each bank of the "boxer" motor and the area cylinder filling size and oil cooler are greater than for a Porsche touring car of this cylinder capacity. Power is taken through a single dry plate clutch to a five-speed gear box with Porsche synchromesh on all gears. The drive shafts to the rear wheels are solid and are unusual in that they have joggled Hooke-type universal joints inboard which avoids the use of splines on the shaft (which can cause trouble) and they accommodate variations in shaft length due to suspension movements.

The car is particularly interesting as it is a very potent machine which is available to the keen amateur driver who is able to indulge in the racing game.

The first 904 to reach this country was delivered some little time ago to that well-known and highly experienced Frazer Nash and Porsche driver, Dickie Stoop.

Details of the front and rear suspension reveal heritage from the grand prix car. At the rear joggled Hooke-type universal joints take up changes in length in camber alteration

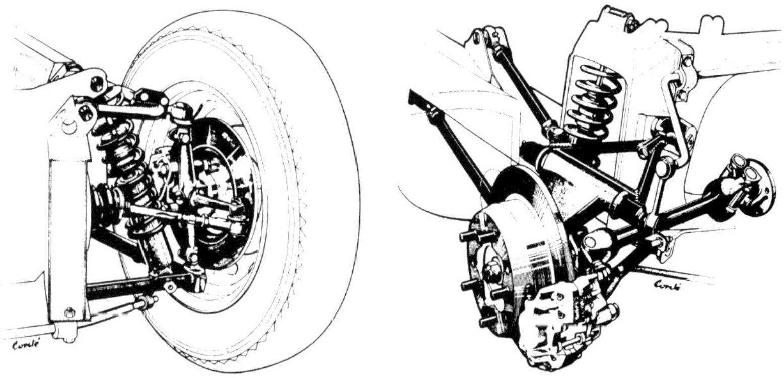

Nearing the ultimate in low build with a penetrating nose, the Porsche still offers good comfort and headroom for the driver despite its overall height of $41\frac{1}{2}$ inches

PORSCHE CARRERA 6

A new racing car from Porsche

PORSCHE has introduced its latest racing weapon; the Porsche Carrera 6. At least 50 examples of the new model will be produced in order to qualify it for the sports car category (Group 4) under the revised Appendix J regulations governing international competition.

The new Carrera 6 uses the 6-cyl, 2-liter engine based on the production 6 used in the Porsche 911, but tuned to put out 210 bhp (DIN) at 8000 rpm. The engine is located ahead of the rear axle and drives the rear wheels through a 5-speed gearbox. A space-type frame built up of small diameter tubes is used and the bodywork is fiberglass. The doors are of the gullwing variety popularized on the Mercedes 300-SL. The overall length of the Carrera 6 is 161.9 in. It is 38.6 in. in height and the factory reports the dry weight to be about 1265 lb.

In conformance with usual Porsche practice, the cars will be offered for sale for customers to campaign. Deliveries are expected to start in early spring and it is anticipated that the required 50 cars will be produced early enough to make it eligible for events counting toward the International Manufacturers' Championship. The price announced by the factory is DM 45,000, about $11,250.

THE CHALLENGE TO FERRARI

Porsche Carrera 6/Ford GT Mark II

PHOTOS ALICE BIXLER

THE INTERNATIONAL CHAMPIONSHIP that is being battled over by the greatest names in road racing is this year known as the "International Trophies for Prototype Sports Cars." This is the big one, the one that Ferrari, Ford and Porsche will be contesting in 1966.

The regulations for this championship specify that there will be two displacement classes— over and under 2 liters. Last year Ferrari won the over-2-liter championship, humiliating Ford at Le Mans and winning where it was necessary at the other championship races. In the under-2-liter class, Porsche had it pretty much its own way.

This year, however, Ferrari is preparing a serious challenger for the smaller division as well (see page 74) and to meet this threat, Porsche has introduced its new Carrera 6. This car is all new, not a development of the Porsche 904 which will now run in the Sports 50 class for the other major championship, the International Championship of Sports Car Manufacturers.

The Carrera 6 has a space type tubular frame, fiberglass body and gull-wing-style doors. Typical of Porsche's appreciation of the necessity for driver comfort in long-distance events, the driving compartment is comfortably outfitted and both the steering wheel and the seat are adjustable.

The Carrera 6 has a flat-6 aircooled engine with a bore and stroke of 66 x 80 mm, for a total displacement of 1991 cc. More than 230 bhp is being developed by this engine (the factory gives the figure as 210 DIN hp at 8000 rpm) and there is usable power available between about 3500 and 8200 rpm, a usefully wide power band. A 5-speed gearbox is used and the top speed of the Carrera 6 is given by the manufacturer as 280 kph, roughly 175 mph. The weight of the car without fuel but with oil is about 1275 lb.

The Carrera 6 was very impressive in its first competition appearance, the Daytona 24-hr race. Driven by factory pros Hans Herrmann and Herbert Linge, the Carrera 6 won the 2-liter Sports-Prototype class and finished 6th overall. At Daytona, however, there were none of the new Ferrari Dino 206/S on hand and when these meet in future races, the struggle should be a classic one.

In the big-engined class, Ferrari's number one challenger is the Ford GT Mark II. This has been developed from the 427-cu-in Ford GT that was rushed to completion for last year's Le Mans race and failed after leading impressively

THE CHALLENGE TO FERRARI

during the opening hours. The Mark II, powered by the huge 7-liter push-rod Ford engine, has been de-tuned slightly to give "only" about 465-475 bhp. This power, combined with the aerodynamic shape, should propel it along Mulsanne straight at Le Mans at something over 200 mph. Last year's version, less clean aerodynamically, was clocked at 185 consistently and was reported to have touched 210 on occasion. The 427 GTs went out with gearbox trouble last year but this component has been thoroughly tested in the interim and should be ready to go the distance.

There will be a total of eight events counting toward the Sports Prototype championship in 1966. By definition of the Federation Internationale de l'Automobile, races counting toward the over-2-liter championship can be no shorter than 1000 km (621 mi) long or 6 hours duration. Under-2-liter championship races can be half that distance, but as most of the races are for both divisions, the smaller engined cars will be covering roughly the same distance as their larger-engined counterparts.

Counting toward the championships in 1966 are the following races:

Feb. 5-6 24-hr Daytona Continental
March 26 12-hr race, Sebring, Fla.
April 25 1000 kms, Monza Italy
May 8 Targa Florio, Sicily
May 22 Spa GP, Belgium
June 5 1000-km race, Nurburgring, Germany
June 18-19 24-hr race, Le Mans, France
Aug. 14 Hockenheim GP, Germany (2-liter division)
Oct. 2 12-hr race, Reims, France

It is not yet known which of the European races, besides Le Mans, will be entered by Ford. Having already won at Daytona, and expected to make a strong show at Sebring, Ford will have to run (and win) at least three European races in addition to Le Mans if it expects to wrest the championship away from Ferrari. There is little doubt, however, that there is more publicity benefit to be gained by a victory at Le Mans than by backing into the championship by piling up points from lesser wins and there is no doubt at all that Ford's major effort in 1966 will be to win at Le Mans.

ILLUSTRATIONS BY

WERNER BÜHRER

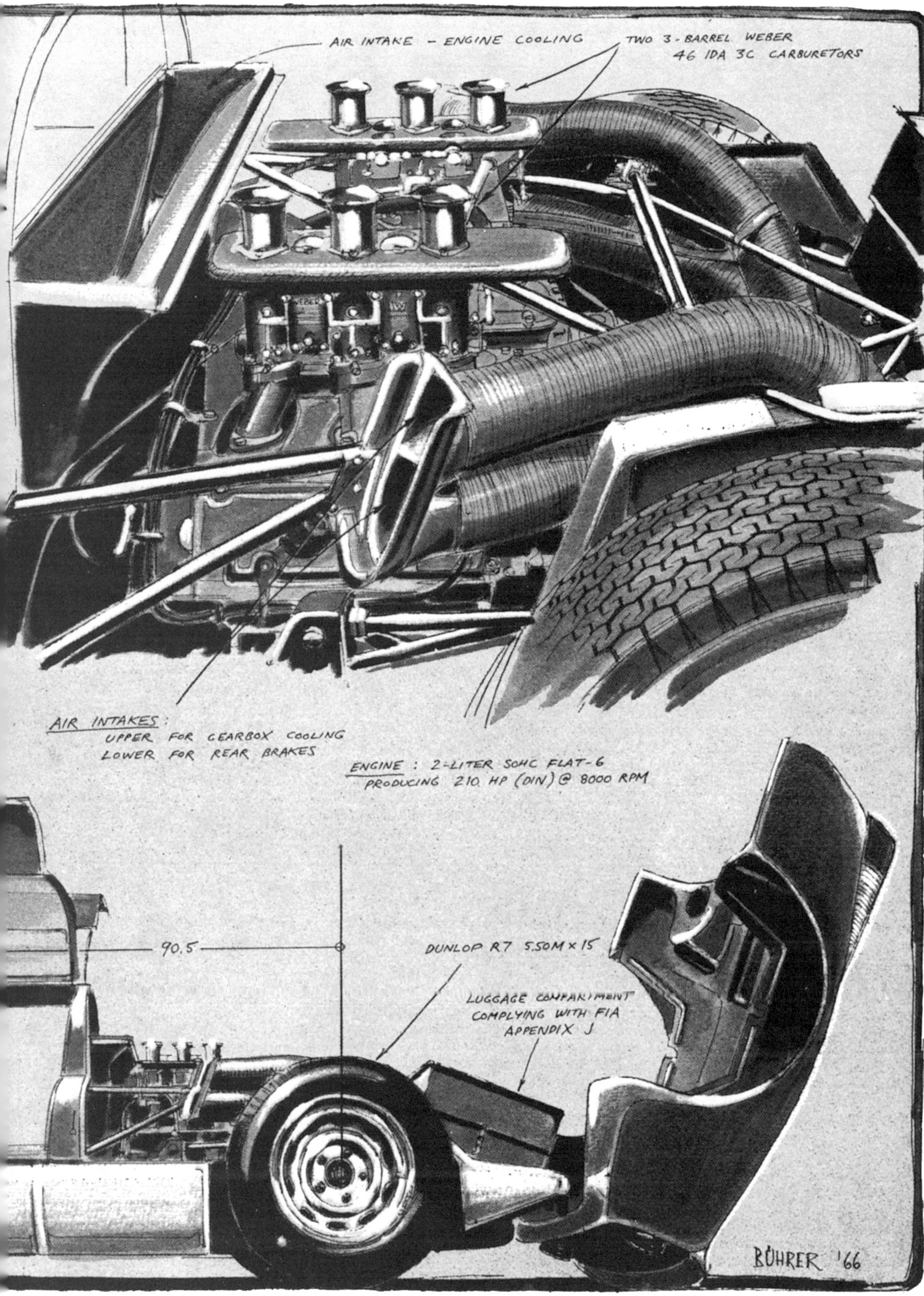
AIR INTAKE - ENGINE COOLING
TWO 3-BARREL WEBER 46 IDA 3C CARBURETORS
AIR INTAKES:
UPPER FOR GEARBOX COOLING
LOWER FOR REAR BRAKES
ENGINE: 2-LITER SOHC FLAT-6 PRODUCING 210 HP (DIN) @ 8000 RPM
90.5
DUNLOP R7 5.50M x 15
LUGGAGE COMPARTMENT COMPLYING WITH FIA APPENDIX J
BÜHRER '66

By Jerry Titus

PORSCHE SIX

The Porsche Six engine is proving its point by winning!

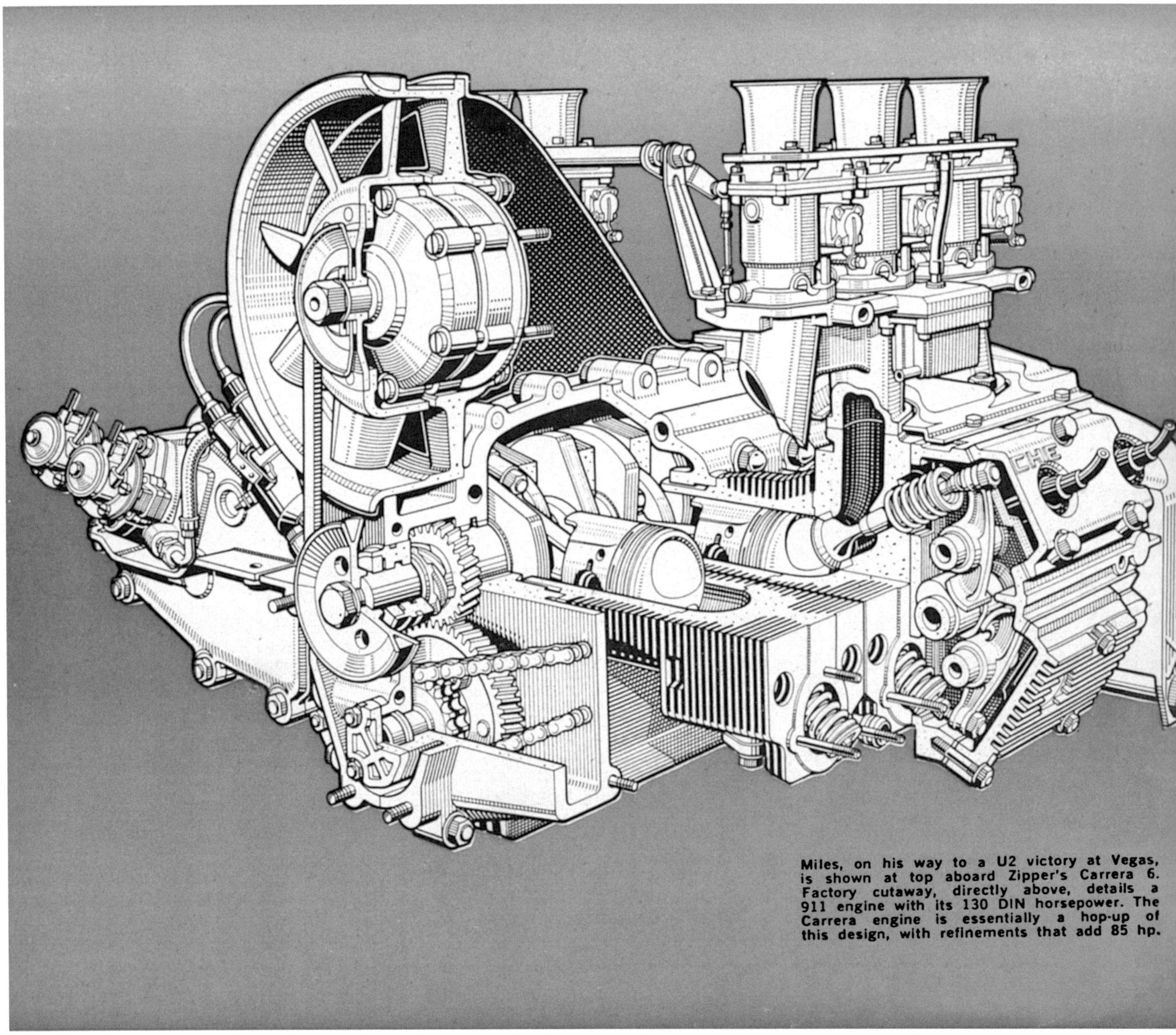

Miles, on his way to a U2 victory at Vegas, is shown at top aboard Zipper's Carrera 6. Factory cutaway, directly above, details a 911 engine with its 130 DIN horsepower. The Carrera engine is essentially a hop-up of this design, with refinements that add 85 hp.

Photos: Bob D'Olivo & Pat Brollier

AS DETAILED IN THE DECEMBER '65 ISSUE OF SCG, Porsche is in the six-cylinder business to stay. It was pretty obvious from first glance that the Carrera Six, announced at the Frankfurt Show last Fall, was going to be a winner. In projecting the SCCA Production Classifications for this year, in the March issue, we predicted the 911 would be a "sleeper" in Class D, not knowing how intimately we would become involved in proving that point. As we go to press, the Carrera Six has just won the Targa Florio *overall*, and thus pretty much cinched the Manufacturer's Championship before LeMans, based on its wins at Daytona, Sebring, Monza, and the Targa. We've been out with the 911 in three SCCA Regional races so far, had no trouble winning the class, and have exceeded known lap times for DP by a comfortable margin. In fairness, SCCA classifies Production sports cars by performance potential, rather than by engine displacement, and tends to downgrade a new model its first year in existence. The car *is* heavy, with an official weight of 2177 pounds, meaning that the legal five percent reduction can bring it to 2068 as it sits on the grid. But we're getting ahead of our story . . .

The debut of the Carrera Six took place at Daytona. As we dutifully reported, it was impressive, but there were no Dino Ferraris there to spoil the romp. At Sebring, there *was* a Dino, driven by a very quick duo. It led, but at a pace Porsche team manager Von Hanstein figured would foster its demise before the end of 12 hours. The figuring was a little edgy, as it lasted to near the 10-hour point before gearbox problems put it out. Carreras were there in force and, when they lost their "rabbit," they had another car in a challenging spot. When *it* crashed, there was still *another* to win. Inconclusive results? Maybe. But Porsche didn't think so and, on the basis of the Daytona results, they were cranking them off the assembly line.

Meanwhile, back at the ranch (Los Angeles, in this case), a dealer named Vasek Polak has already visited the factory, and turned down the Carrera Six. First, he felt it was too heavy at 1325 pounds, for the relatively sprint-type races of the USRRC and US pro series. Second, many of the races held in the Far West are run in ambients well above 100 degrees, so a coupe configuration can be a problem. But, he managed to purchase one of the pre-production Carrera Six engines, and a development chassis — basically Elva, but *much*-modified for six- and eight-cylinder tests, and hillclimb competition, by factory driver Gerhardt Mitter.

Another dealer, Otto Zipper (Zipper-Estes Motors and Precision Motors, of Beverly Hills), decided these were not drawbacks for his purposes and ordered two Carreras for his chauffeurs, Ken Miles and Scooter Patrick. One car, now partially sponsored by Russkit, and duplicated in slot-car form, was delivered in Mid-March, the other some weeks after Sebring. Polak's car arrived in mid-December but, due to the press of business and the opening of a second store, it was mid-April before he managed to install the engine. This writer soon got a first-hand taste of what it is like to have the new gearbox and powerplant propelling him.

As of last Fall, we wound up owning the under-two-liter Modified record at Riverside, aboard Vasek's 1800-cc Elva-Porsche. It was a record we were proud of, because quite a few first-rate under-twos had taken a crack at it. The time was one minute and 34.6 seconds. Hugh Dibley of England was the prior owner, aboard a BT8 Brabham, at 1:36 point-something. With the new and unsorted chassis, old-style tires, and the wrong gear ratio, we soon managed a 1:34.0 and could realistically see another couple of seconds. The power flow was smooth as glass, from 5500 rpm to the 8000 rpm redline. Engine noise was very low, the exhaust note well behind us, quiet but very Ferrari-like. This was a surprise, because from trackside the Carrera Six emits a distinct bellow from the twin exhausts that stands out in a field of howling big-bore machines. Miles, testing the Carrera, was understandably reluctant to make his lap time known but, when Ford Motors decided they needed his services the weekend of the Riverside USRRC, Dave Jordan took over the ride and smashed our record officially at 1:34.5, and confided that he felt there was a reasonable margin for improvement left.

The five-speed transaxle is one of the best we've ever handled. The ratios are close, but not too close, and the short, firm gearshift gives you a real feeling of security as you move through the gears. The prospect of missing a shift and busting an over-$5500 powerplant is not joyful, and anything that minimizes this possibility is appreciated, even though you've heard that the factory test engineers ran one up to 12,000 rpm for a short burst without blowing it.

PORSCHE SIX

The main differences between the Carrera engine and the 911 are in the use of exotic metals in the former — titanium connecting rods, magnesium crankcase, alloy rocker arms, and so on. The pistons are domed, and the cylinder head is drilled to accept a second sparkplug for dual ignition. The camshafts are, of course, wilder, and valve adjustment is accomplished by shimmed caps atop the valve stems, instead of adjustable rocker screws as in the 911. The flywheel has been Swiss-cheesed to reduce its weight, and the clutch disc is sintered metal.

About the same time Vasek started putting the "Porsche 6 Special" together, he decided that it was high time someone started racing a 911 in SCCA National competition. A white coupe was pulled off the showroom floor and stripped out, the engine disassembled, and sent out for balancing and Magniflux. American Mag wheels were purchased, with the proper offset and a 5½-inch rim width. Goodyear Blue Streaks were installed. Seat belts, rollbar, and bucket seats took care of the interior, and both side and rear glass was replaced with plexi. Aside from Koni shocks, the suspension was left standard, though we intend to at least roll-stiffen it some more.

Polak is secretive about what he did to the engine but, from what we occasionally observed as it went together, it has domed pistons, enlarged ports, slight combustion chamber modifications, and slightly enlarged venturis in the Solex carburetors. By the time you read this, all 911's will come from the factory with Weber carburetors as standard. This will allow easier jet calibration, but otherwise won't make much difference.

The clutch was a weak point that came to light during our first practice session with the car at Las Vegas. The lining pulled loose from the disc, giving no warning of slippage beforehand. A heavy-duty disc seems to have cured the problem, even though we get some strange odors in the

Continued on page 173

Temporary body modifications, above, were by a designer who shall remain nameless. Nose is now a foot shorter on the Porsche 6 Spl. and a complete aerodynamic body is planned. The Carrera engine installed in the Special requires a wide bay and increased side-rail strength. These modifications were done at Porsche factory when it became mobile test bed for both six and eight-cylinder engines.

Above, the 911 with stock springs and sway-bar is a bit too "soft" in the corners, but still very controllable. Below, the interior was stripped and racing bucket seats installed to bring it very near minimum weight. At right, the 911 Six is redlined at 7400 rpm and is probably putting out 185 DIN horsepower, but Polak built the entire car in two weeks and hasn't yet had time to dyno-check.

Top, Patrick takes U2 win at Riverside USRRC race. Middle right, front discs of Carrera bolt to six-stud alloy hub. Fifteen-inch wheels are used to provide maximum air circulation around the brakes. Far right, Miles on his way to Laguna U2 victory. Below, engine compartment of the Carrera is extremely sanitary and uncomplicated. Upper ducting brings cooling air to transaxle via fender scoop.

DRIVING PORSCHE'S PANCAKE

The car that rubbed dirt in Ferrari's wounds at Le Mans is a docile darling to drive, says our man Jerry SLONIGER. (He tried it at Hockenheim!)

THE OVERHEAD DOORS ON Porsche's latest pancake, the Carrera 6, remind me very strongly of that Ustinov spoof about the Schnorcedes team and their drivers in a hurdling position for proper weight distribution. As his take-off on the dreaded Moss explained in a bemused tone, 'they forget that hurdlers don't stay that way'.

You wonder if it has occurred to the Porsche minions – young Ferry Porsche III, who designed their current 911/912 road cars and the 904 track machine, is quick to deny this one – that a super-light glassfibre door with three catches may be fabulous wear on the scales but is hardly the thing for Le Mans starts, those classics of confusions so dear to the hearts of long-distance race organizers. Still, they got through the 24 hours all of a piece so maybe Porsche drivers are trained to latch-up . . .

The Carrera 6 is certainly no harder to eel into than any modern quasi-two seater, though it will do nothing to abate the search for jockey-size adults who can drive. But in other ways it does seem a pendulum return from the 904, which was admittedly sturdier than a two-litre racing car need be. When that highly successful model was announced Porsche made quite a point of 'building a car for real long-distance races, not artificial sprints' – and were good as their word. Look at the Monte.

Comments of this order are conspicuously absent where the Carrera 6 is concerned. The new space-frame car (dropping the 904's idea of a backbone and bonded body in the interests of easy repair, among others) is credited with only 1270lb dry. To do that in a genuine two-seater coupe means that non-stressed panels like the gull-wing doors have all the rigidity of starched linen.

Due to their size and overhead hinges these doors are unhappily prone to bell out in the wind like spinnakers if not hooked in the middle bottom by a snap catch and around the edges by bolts, much like the ones on our chicken coop back home. For trundling journalists about or even giving your few favourite friends a short run this is fine. They can do up the doors to gloss over that old nervous twitch which comes from starting an unfamiliar full-blooded racing car in front of all those works drivers – people who very seldom stall it three times.

But the operation of a lock system on a par with Fort Knox is almost too much to ask of some poor driver desperately trying to get away under the eyes of most of central France and half the world when the flag drops at Le Mans. The very first run during Porsche's recent demonstration day at the new Hockenheim course made this point all too clearly when the wind got under an improperly latched door and took its window clean out . . .

Running about in the doorless car later (only the window actually blew out, but the wind-tunnel effect would have finished the job if they hadn't unscrewed the lids completely) proved at least that the curved windscreen does its work. Mind you, it was a mite draughty down the straights but you weren't really buffeted at high speeds until you tried to help out the cornering in approved sidecar manner and stuck your neck and/or head out.

Regarding the car itself, it might be added that it really is a two-man effort if they ever get mad about rallying. In fact the passenger footwell is so long you disappear from sight like a rabbit spotting a hawk every time the driver brakes hard. This point is worth mentioning since so many so-called sports cars leave the passenger's side fully prepared for Toulouse-Lautrec. The driver in the Porsche has ample room to be a layabout too, provided he doesn't really want to read the rev counter in a rush. Talk about shrouding your instruments to avoid reflections! The tachometer is so deeply recessed you have a strong inclination to nudge it with one toe.

So is the oil temperature dial, for that matter. That item, reading both F and C on one dial, is angled at the driver from the middle but so far down a well that one Porsche works man who shall remain nameless hadn't the faintest idea what it was. You could get a bad case of tunnel-vision reading it to the nearest degree – not that Porsches are prone to run all that hot anyway. At least it won't reflect.

In a way this is a report on a car which is no longer produced. Even if you like the idea, have the 4100 or so quid (tax extra, if any) and sufficient standing – an esoteric but vital part of any Porsche transaction – you still can't own one. The current FIA rules for sports cars specify 50 examples before you can be

Look, mum – no doors! Window blew out at speed during our man Sloniger's testing, so car spent the rest of the day without its gullwing flaps. At far left, works driver Mitter plays the Michelin man as he hurtles down Hockenheim's long straight at 140. Below: at least it makes a passenger feel better in the corners . . . Above right: doorless or not, Carrera 6 looks the part while Mitter just looks jealous as he surveys sister car which kept its entry-hatches intact. Entry isn't easy either way, although driving position (demonstrated, below right, by Hans Herrmann) is quite comfy. For long-distance racing each driver has his own seat-insert, moulded to fit. Neither of the cars shown here had the latest aerodynamic spoilers below headlamps, or the special long tails fitted to some for Le Mans

homologated. Porsche took no chances with the Federation's arithmetic and built a full 60 – all of which were bespoke long before they were finished.

If the Carreras continue to hurry home in front of Enzo Ferrari's new V6 Dinos (no matter by how much) the company will have no trouble selling twice or three times as many for the price, but every one of the original 50 was individually test-driven at Hockenheim, hand-assembled of course and variously pampered, so that to build more couldn't really be a commercial venture unless they wrote them off to the ballyhoo budget.

Getting down to cases, the car you can't have unless your name is d'Udy boasts a very tuned version of the production two-litre flat six with one overhead camshaft per bank – the latest Porsche powerplant, in effect, and one of the company's best compromises between up-to-date design and reasonable service. The old Carrera four with its four overhead cams was a nightmare for plug changes and similar simple service, while the new one is merely a racing engine of unusual robustness.

It punches out 210bhp DIN at 8000rpm with 145lb/ft of torque. The two triple-throat Webers fitted at present are only an hors d'oeuvre: Porsche men are playing with fuel injection too. Questions are apt to be answered with profundities like 'I don't know whether we can talk about that', so we won't expect it in the production cars this month, anyway.

Torsion bars may do the job for the road-going 911/2 but the Carrera 6 uses coils front and rear. It feels remarkably solid and non-twitch, rain or shine, and we had both at Hockenheim. Like most modern sports-racers the white Porsches take even the fast sweeping corners flat and unflurried. We didn't realise how clinging the demonstration pair could be (Dunlop Yellow Spots on one, Green on the other) until we did a few laps trying to hold one of the 911 rally cars in something approximating a straight line at even a modest (for them) 125 mph with only very gentle winds.

Admittedly this particular 911 was the Klass Monte car and looked to be sitting on stilts to get enough ground clearance, which obviously did its straight-line tracking no good at all, but as colleague Manney commented: 'If this was the only Porsche a man drove they wouldn't sell many'. With ultra-close rally gearing the 911 needed plenty of road for us amateurs at anything over the ton – and we vote Herr Klass twice as brave as we thought he was after watching him rally.

Compared with this extremely specialised 'production' product the Carrera 6 felt safe as houses, going a fourth as fast again to boot. Nobody could call the modern Porsche a dream car for looks, but it certainly fits the prevailing racing scenery, which is all knee-high, too.

Carrera 6 racers have matched the monsters everywhere from Daytona and Le Mans to the green hills of old England, and frankly we are agog and full of anticipation too about their debut on the European hillclimb scene. After all, their main rival down in Maranello has to find 50 handsful of large-denomination lire first – while the busy elves of Zuffenhausen are ready, tested and (as usual) overbearingly confident.

For a pleasant change, this one is all go and no show

IT'S those little detail touches which make a Porsche. When I first tried the Carrera Six I lacked the proper perspective of trying one on a test track — like those door catches to make the Le Mans start such a nightmare. As the dreaded Moss might have said (he didn't really; it was that wonderful Ustinov parody — but it fits): the Germans sometimes get so efficient they overlook efficiency.

A goodly number of Porsche 904 drivers have complained about the heft of that first fibreglass Porsche — some even fitted alloy tail pieces and saved 50 to 80 lb without even trying. As Porsche itself noted when the 904 was introduced, this is not a sprinter but a proper, traditional race car for the Targa and its peers. Such comments are conspicuously absent with the Carrera 6.

The new design — which Porsche's resident felt-pen man named Ferry Porsche III who drew the 911/912 and 904 is quick to deny any hand in — is undeniably smaller in frontal area, faster and lighter. After all, if you use non-stressed parts like doors which have all the rigidity of this page you expect to save the odd ounce.

The whole theory of a Le Mans start, on the other hand, requires your keyed-up pilot to get

PORSCHE CARRERA: THE UGLY

The Carrera 6 prototype. Subsequent models haven't got any prettier, but they have got faster and more aerodynamic.

Our man on a terror test: Sloniger tries out the Carrera from the passenger's seat after the door blew out in an early run.

underway ahead of the snarling hordes. Some hopefuls to the contrary, it isn't your speed in the 20 yard dash which counts but rather your ability to eel in, batten down all hatches, find the switches and move off (take that gear lever out of your trouser leg, Juan-Manuel).

The annual panic of Le Mans apart, I wouldn't really reckon a rapid start which included opening a door so flexible it will bind if 1/16 in. out of true on opening and so casual about its closing that you need not only a snap catch at the bottom but bolts at the corners like our chicken coop back home. I admit that they managed very nicely at Sebring, but what about the major leagues?

Three-handed drivers at the Nurburgring 1000 kilometers can always shoot their bolts down the counter straight but Le Mans doesn't have any really straight bits until you hit Mulsanne and speeds which will bell out any Carrera 6 door not battened down at every conceivable spot like a spinnaker in a hurricane.

When the German firm gave some of us a ride on the new Hockenheim track in two of its toys — with special drives for the usual favorites — lap one produced the first improperly fastened door and the first window blown out — and that's a slow course. From then on we drove about *al fresco* since the only way to keep the feather weight doors from going too was to leave them in the pits.

Hurrying down the back dog-leg you soon found that the windscreen it more than expected

CHAMPION

Someone even dubbed it the Zuffenhausen Zot

Carrera predecessor also won (GTS) first laurels at Targa, and later excelled in all types of road races.

By SLONIGER

Inside, the room isn't exactly generous for a big man, but jockeys are supposed to stay slim for 2-litre racing.

and provided you didn't stick your neck out, with head attached, the buffeting was less than violent — of course nobody had the wheel for 24 hours. Mind you it got a mite draughty down the straights but the streamlining is better than you think.

I'll hardly be asked to complete the team for Le Mans anyway, so we can proceed to the car as mass transport and forget about Le Mans starts except as material for small side wagers.

Curiously enough in a highly specialised sports car Porsche has fitted two proper seats and even added sufficient foot space for the intrepid passenger. *(FIA regulations—Ed.)* In fact the passenger foot well is so deep the unwary tend to disappear from sight like rabbits sighting a hawk every time the brakes are used hard. We all said no sane man would rally the 904 either and look what *they* did in the Monte, so perhaps Porsche has even madder ideas handy — after they build the second batch. They only need 940 more to qualify for GT, having completed the vital first 60.

Actually sports car rules, '66 style, only require 50 but Porsche knows all about the FIA system of arithmetic and threw in 10 extras for luck. When you come right down to it I am dithering on about an out-of-production automobile here since no amount of money will buy you a Carrera Six. At something over 10,000 of your dollars, ex works, Porsche had far more orders than potential Carreras even before it became a back-issue item. Not even that vital special standing which counts for more than your beer drinking ability at Porsche can buy one now.

For that matter the factory could doubtless sell two or three times as many if there were anything like a commercial proposition. For such a small factory, 60 is already a considerable load and a praiseworthy one since it could gain nearly as much ballyhoo for the same money racing all alone, within the family.

Customer racers have always been a Porsche tradition — with some even getting nearly as fast as the "identical" factory cars — and they decided to keep it up. Each and every Carrera Six was not only hand-assembled but also hand-tested at Hockenheim to ensure smooth operation of those linkages, bits and six door bolts.

Getting down to cases most briefly: the cars we sampled were producing 210 (DIN) bhp at 8000 from the flat-six single-OHC engines and sitting on the road like the proverbial white stripe. In fact the handling is so good you have to drive a factory 911 rally weapon for comparison. Porsche happened to have the Klass Monte GT winner along, featuring a very slightly tuned engine within the current strict GT rules and a body sitting up in the air for rut clearance.

This daddy-long-legs posture did nothing for the 911's twitch border and even with ultra-short rally ratios which only gave an indicated 125 mph at 6900 in top (fifth) gear, the car was all we amateurs could manage down the straight and it wasn't even windy. I suddenly realise how brave Klass can be. Then you get into the lawn chair Carrera Six and go into even the mildest of orbits: night and day.

Needless to say such Hockenheim efforts were conducted at very modest rpm by racing standards, while the Carrera Sixes looked far less sanitary at the recent Le Mans test day. They all flew nose up except the quickest one: which is why Porsche goes to Le Mans 10 weeks before the race — to sort out handling. Incidentally, the factory boys were playing with injection in place of two triple Webers then too but nobody wanted to know about that during the tests at Hockenheim.

Driver-wise the Carrera Six is simple to the point of starkness. Directly in front of the driver is buried a tachometer so far down a matt-black well you have an odd feeling you can nudge with one toe. It isn't a quick-reader but it doesn't reflect on the glass either. The other instrument proved upon hard peering to be the oil temperature, down a different mine shaft. Lest you think I exaggerate the depth of shrouding: one of the two factory aces (and the precise one shall be nameless) didn't know what was down that second tunnel himself: too busy going fast.

The Carrera Six has already swept its class in most types of road racing in the Continent and the US. If Enzo Ferrari persists in his solitaire game of one Dino per race Porsche will be hard put to lose so much as a single event towards the manufacturers' championship. Should the V6 Dinos appear in strength we can expect a battle royal. The busy elves of Zuffenhausen are tested, proven and as usual overbearingly confident. Of course they have a product to boast about. #

First race for the Carrera coupe was at the 1966 Daytona Continental 24 hours race where the cars again dominated their class

Porsche trio, again at Le Mans. De Klerk/Schutz, Hermann/Linge, and Klass/Stommelen thunder through Mulsanne.

Leading the way to a Lola 70, is South African Peter de Klerk's Porsche Carrera. This is at Oulton Park where the Carreras won the 2 litre class in the Gold Cup.

Two cars, showing the constant-curve of the cockpit down to the Kamm-tail, and the abrupt nose.

Carrera 6

PORSCHE has finally entered into the big gun racing classes with the exciting Carrera 6. This car made its debut at the 24 hour Daytona race and made quite a showing for itself. It is aerodynamically designed and the body areas separate so that the engine can be worked on quickly in an emergency. The power is supplied by a 6 cylinder 200-210 hp engine, and is capable of a top speed around 175 mph. We hope that these pictures will give our readers an insight into this fine piece of racing machinery.

MORE

Carrera 6

Carrera 6 with doors open gives the illusion that it is ready to take off. Scoops in front are used to circulate cool air through cockpit. The head and running lights are enclosed in clear plexiglass.

Set up for competition the Carrera 6 leans into a turn during pre-race practice. In this form the machine is capable of speeds up to 175 mph and it runs in the under 2 litre class.

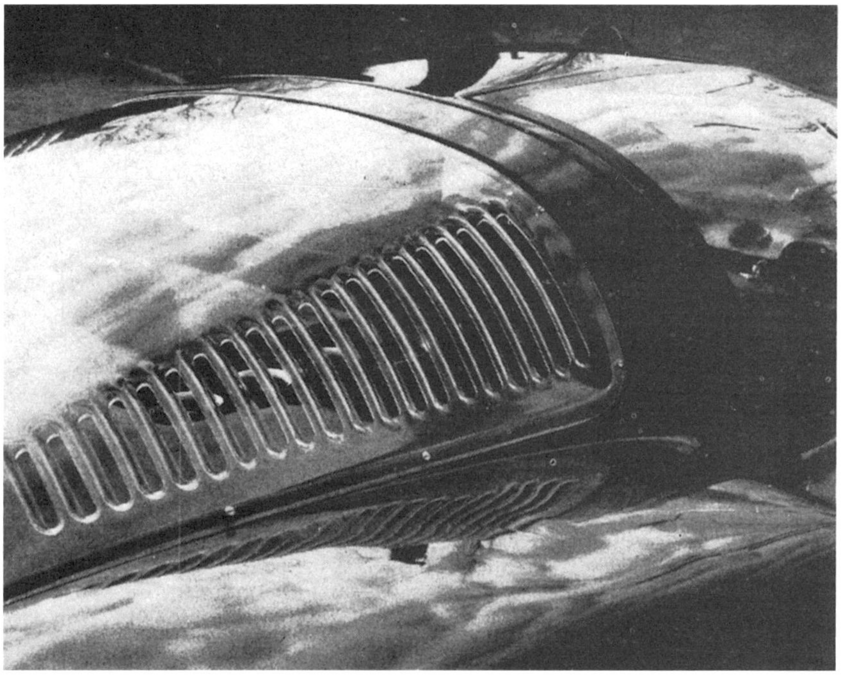

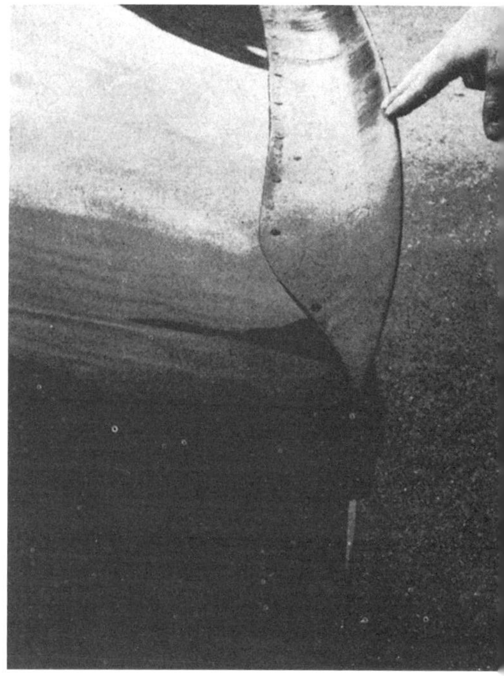

Top left — Louvers in the rear windshield allow hot air to escape during races. The windows in the Carrera 6 are plexiglass and very light but durable.

Above — The upswept fin on the rear of the car gives more roadability and stability during the high speeds reached during competition. Many of the sports cars have adopted this idea which was started by the Chaparrals of Jim Hall.

Left — Porsche driver peeks out of the little machine. The size can be estimated from comparison with the larger cars in the rear.

Below — Profile view gives the impression of speed. The sleek lines of the Carrera 6 contributes greatly to its impressive list of victories.

Manney drives the PORSCHE CARRERA 6

The best value-for-money racer on the market today

BY HENRY N. MANNEY

Ah it is a truly wonderful spur that competition brings. Consider for example the lowly Volkswagen, motoring equivalent of Salvation Army lassies' shoes, that was seized upon by Porsche as basis for a high performance car. Early doubts about excessive oversteer were resolved into later doubts about excessive oversteer but a series of clever factory drivers like the late Edy Barth and Herbert Linge ensured that on the competition cars at least it didn't get out of hand—or at least they had the knack. These racers ran from the early tinfoil Glocklers through the shapely 550s, RSKs of various sorts, and the unsuccessful Formula 1s to the Targa-winning 8-cyl roadsters. On the GT front, Porsche didn't really go to town until the Abarth-Carrera came out followed by the fiberglass 904, which car confounded the experts by doing well in rallies and even, in the hands of the underrated Ben Pon, blowing off much bigger cars on high speed tracks. Came the new FIA regs already and everyone wondered what Porsche would do—*ach du lieber,* it's Batman to the rescue!

The general layout and recent history of the Fledermausmannewagen has already been covered in a recent issue so there is no need to do more than point out that with it Gerhard Mitter recently carried off the European Hillclimb championship and in the classic road races it won the 2-liter Prototype championship ahead of Ferrari's 2-liter Dino.

Now we had been after press relations boss Von Hanstein for months to have a drive in a Carrera 6 but all resources were being strained to sort out the prototype while still building cars for the team and customers. Consequently it was with some joy that we received a summons to report at Hockenheim track. This track, as you may have read in these pages, has been modified from the old egglike oval due to an Autobahn coming across the bottom end, and now boasts great concrete grandstands surrounding a network of club tracks in a bowl while the previous long curve and return straight are attached thereto. Much shifting of sand is going on and it will make a first class venue for sports and F2–F3 racing in that part of Germany, if a little small for F1.

Naturally enough it was raining when we arrived and naturally, racing cars being the uncertain things they are, nothing approaching a competition machine was to be seen. Just drivers Mitter and Hans Herrmann, a couple of German journalists, and a spaniel wearing a Tyrolean hat and smoking a pipe. Eventually various odds and bods began to show up, enough to know that we were in the right place anyway, but no racing car so we amused ourselves taking laps around in one of the normal Porsche 911s to see where the track went. Everything was quite wet and we didn't press our luck, not wishing to find out about rear-engined cars especially when Von Hanstein was around—not that it didn't feel quite sanitary. Anyway sooner or later one of the Batmobiles blubbered up, a brand new one as it turned out, and with some difficulty a rather corpulent English journalist was inserted into the passenger's seat. It was even more of a project to get the door closed as it requires three catches and sundry bangs but off Mitter went with his passenger looking rather like those herring you see coiled up in glass jars. A lap later they were back again as the rather feeble fiberglass door had blown off, probably as a result of both occupants breathing in unison, giving the aforementioned rcEj the fright of his life. As by now things were supposed to be well under way, Von Hanstein was in a parlous state running around flapping his arms like a rooster but the mechanics soon had the doors off (or remnants thereof) and the Carrera 6 set forth once more as a species of Jolly or beach car.

Eventually the other one arrived as it had been held up at the factory for some unknown reason; Herrmann plumped himself into this and as I was next in line, I went with him. It is a bit of a problem getting in over the wide sill and down into the hole, especially as the floor-mounted fire extinguisher lies in wait to snare your trouser leg and you don't feel that you can really put much strain on most of the upperworks. Once actually inside, there really doesn't seem to be room for two people with shoulders, rather as on some of those economy charter flights, and the best thing to do is sit sort of catty cornered so that the driver has room to operate. This Hansi did in exemplary fashion albeit poking the clutch pedal from time to time as he imagined a bit of slip. There did seem to be an odd vibration around 5500–6000 so after I got out it was taken away and worked on for a while. As far

Passenger had more headroom after door came off at speed.

JERRY SLONIGER PHOTO

PORSCHE CARRERA 6

as the ride was concerned it was very pleasant as Hans is quite safe even in the streaming wet and the car didn't feel the least bit twitchy. My only complaint was that water came in through the body joint and dripped on my ankle.

In the interim before driving it myself, we took out one of the rally Porsches that suddenly appeared, Gunther Klass' car as it turned out. This was set up rather high and with very short gears (all normal enough for rally cars) but what scunnered us was that it felt absolutely eetery-teetery all around the circuit. Klass must be the King of Oversteer or else deadly sensitive as that one would be too goosey for me in the dry, unlike the normal road-going one.

Anyway soon after that Von Hanstein let me know that it was time for me to do the do or die bit so I crawled into the driver's side of the Jolly alongside an apprehensive Mitter, who straightaway told me not to use fourth as it was jumping out. That was a lot of help as I have enough trouble remembering where the gears are on a 5-speed Porsche box let alone avoiding fourth. Crowding into a borrowed crash hat, I then fired the thing up and after an ineffectual glance over my shoulder trying to see through the body pillars and yellow plastic back window, progressed out onto the track with only a mild bit of ugga ugga.

The engine fitted to this car was redlined at 7000 but nevertheless didn't seem particularly cammy; either a fairly cooking mill had been dropped in for the Gentlemen of the Press or else there is a lot more power to be got out. Driving it was perfectly straightforward as the directional stability is first-class, the view of the road between the Wagnerian-soprano fenders excellent, and the driving position comfortable enough with smallish wheel and tight fitting fiberglass buckets so that it wasn't a tour de force to steer it at all even if the gearshift was a bit close for my liking. There was little to distract the driver inside as the only two instruments fitted that I could see, a small tach and even smaller oil temp gauge, lived down two of the many tunnels (the others are air feeds) with which the matte black dashboard was riddled.

The whole business felt very comfortable in spite of a distracting distortion in the screen corner but just the same it is just as well that a short rear end was fitted as who likes to go steaming along 120-mph corners at 120 mph. As it was, I speedily found myself doing 6500 in fifth along it in the wet which was quite enough, leaving 500 rpm to bong the nose in if something started to slip. Not that it did as the car felt very stable indeed and as all the controls worked smoothly and easily, the Carrera 6 would even be quite practical for road use.

Next time around I got a bit braver, helped by Mitter who was making encouraging signals and demonstrating shift points with forward flicks of the hand, but the Carrera felt just as solid as before with no instability even in the streaming wet. It seems to be a neutral-steering automobile even at the slow speed I was going and as safe as houses for when I got the back out once on a slow corner through an incautious application of boot, a small left lock correction brought it back again. Frankly, it is one of the few competition cars that I felt as if I could drive in competition given a bit of practice; with such comfortable roadholding and a demonstrated turn of speed the Batmobile is probably the best customer's value-for-money racer on the market today.

PORSCHE CARRERA 6 SPECIFICATIONS

List price	$11,250	Top speed, mph	174
Engine	Horizontally-opposed 6, dohc	Frame	tubular
Bore x stroke, mm	80 x 66	Front suspension: diagonal upper and lower arms, coil springs, telescopic shock absorbers, anti-roll bar	
Displacement, cc	1991		
Bhp @ rpm	225 @ 8000		
Torque @ rpm, lb-ft	155 @ 6000	Rear suspension: diagonal upper and lower arms, coil springs, telescopic shock absorbers, anti-roll bar	
Carburetion	two 3V Weber 46 IDA 3C		
Clutch	single dry plate	Dry weight, lb	1430
Transmission	5-speed in unit with differential	Wheelbase, in	90.6
		Track, front–rear	52.6/55.1
Synchromesh	on all 5	Overall length	161.9
Final drive ratio	4.428:1	Width	66.1
Gear ratios	2.643, 1.778, 1.318, 1.000, 0.821:1	Height	38.6
		Frontal area, sq ft	11.7
Brakes	ATE-Dunlop discs	Fuel tank capacity, gal	26.4
Tire size, front/rear	5.50L/5.50M-15	Engine oil capacity, qt	10.5

continued from page **160**

cockpit after leaving the starting line.

Handling has worked quite well, considering the minimum of modifications. The 911 firmly understeers, but it can be tossed a bit and then held there, with the ample power available. It's at the expense, of course, of increased slip angles. We hope to stiffen the chassis to eliminate slight *inside* front wheel-lift in tight turns, and excessive *outside* rear tire temperature at its inboard edge (indicating too much negative camber for ideal tire bite). Braking is excellent – completely stable, with very little wear and no signs of fade. The slippery body shape contributes to the 911's high-speed velocity and stability, but requires that fresh-air ducting both in and out of the cockpit be carefully planned (something we have in the process right now), to keep the driver cool in hot weather.

Fitted with a special muffler, the 911 is extremely quiet for a racing car. This is disconcerting when you fire it up on the grid amidst the thundering herd. Watching the tach is the *only* indication you have of knowing that it's running, and the same is true for the first hundred yards off the start. Shifting by ear and a quick glance at the tach – as you can normally do in a car you're familiar with, and a course you've practiced on – is useless.

Getting back to the Carrera: Miles *really* wiped out another record we held at Laguna, not only in qualifying, but in the race itself – by almost two seconds! Despite a spin and other minor problems, he had no difficulty winning the under-two category from consistent George Follmer in his 904-powered Lotus 23. We'd hoped to compete in this three-race series (Vegas, Riverside, and Laguna) against them, but managed to break a con rod when a bearing seized (the mains are rollers, but the rods have inserts) on our second qualifying lap at Las Vegas. So the 'Porsche Six Special' remained side-lined, awaiting parts through the trio of events. Failure appears to be due to something foreign in the lube system, as all other bearings were perfect, but two or three of the initial development engines were rumored to have been broken during dyno tests – the cure being enlargement of the oil galleries to the crank. Factory power curves have rated their engines as high as 220 DIN horsepower – in the 240 SAE category – and over 175 mph has been recorded with the coupe.

Plans for the cars currently competing in this country show only local competition scheduled for the Zipper cars until the Fall pro series returns. We'll try and talk Vasek into some Eastern events with the Special, and a full slate is already planned for the 911 in SCCA National competition, with the ARRC runoffs at Riverside in mind at year's end. Meanwhile there's no doubt about Porsche's sixbanger being on the scene – both here and abroad – and proving its point by winning!

Track Test/Porsche Carrera 6

By BRIAN SMITH

ONE of the most amazing cars in international racing is the Porsche Carrera 6. With only two litres of engine this super sleek GT always worries the larger capacity cars and on several occasions has beaten them for outright wins in Group 4 races. While in South Africa for the Springbok Series recently, I was able to test Mike de Udy's Carrera which was giving Ford GT40s and a Lola T70 a hard time in the races out there, and it was a very interesting experience.

On looking over the car, it is apparent that the reason for its tremendous performance is low weight. Not that the car is fragile—in fact far from it, as its record in long-distance racing shows, but there is no unnecessary weight at all and with fuel and oil the car scales only 11 cwt. Incidentally, the 2.2-litre eight-cylinder engined version is faster still, but at the moment is confined to Group 6 prototype racing as Porsche have not yet made 50 models to get it homologated as a Group 4 car.

The chassis is a fairly conventional spaceframe, covered with a glassfibre body of ultra-futuristic shape with gull wing doors and very high wheel humps at the front. This makes the car look fast before it even moves! Roll-over bars are incorporated in the windscreen surround and over the engine bulkhead, but these are really too small to prevent the roof collapsing if the car should roll.

The 1991 cc engine is a flat-six with twin-ohc and air-cooling by a fan mounted on an extension of the generator shaft. Both the block and heads are made of magnesium alloy and this plus the lack of cooling water and radiators saves a considerable amount of weight. And of course being a horizontally-opposed engine, what weight there is, is kept very low in the car—always a good thing. With two 46IDA3C Weber triple-choke carbs, the engine gives 225 bhp at 7700 rpm with an excellent power range from 5000 rpm upwards.

The clutch is the Porsche single dry-plate diaphragm type, as used in many other makes of racing engine. Transmission is through a Porsche five-speed gearbox which has synchromesh on all gears (what luxury!) and a ZF limited slip. ZF also supply the rack and pinion steering. Suspension follows current practice with double wishbones, telescopic coil spring/damper units and anti-roll bars, but there are no fewer than 22 greasing points! 15 inch disc wheels are used with alloy rims of 7 inch width at the front and 9 inch at the rear. As standard they are fitted with 5.50 Dunlop R7 yellow spot tyres all round, but in view of latest developments, these widths could probably be altered to advantage. The brake discs are 9 inches front and 9½ inches rear operated by the usual twin hydraulic system.

Electrical features include dual distributors with four sets of contact breakers, twin coils and two plugs per cylinder. Two Bendix electric fuel pumps are used to take fuel from the two side tanks which form the lower half of the body. Another interesting point is that the oil pumped out of the dry sump goes to the oil cooler only if it reaches a certain temperature, otherwise it is by-passed.

The cockpit is very basic, both in finish and equipment. The only instruments are a tachometer red-lined at 8000 rpm and an oil temperature gauge. An oil pressure gauge is evidently not thought necessary by Porsche.

With the gull-wing doors the car is quite easy to get into, but it is most important to fasten the safety catches or the doors fly

open again—a nerve-racking experience! I found the driving position very good, with everything in easy reach, but although I am only 5 ft 10 ins tall I could not sit properly unless I canted my head over at about 45 degrees to the right to avoid my helmet hitting the roof! Different seats are obviously required for taller drivers. The other poor point was the lack of visibility. To the front the wheel humps mask the road to either side, and you cannot use an interior mirror as the carburettor trumpets stick up in the way and the rear window is nearly horizontal and thus opaque! The car was also a left-hand drive model (Porsche do not make a rhd version), so one way and another I was not very much at home in it.

However, after a few laps I was able to settle down somewhat, but I made no effort to really push things, particularly as I was testing the car during official practice for the Pietermaritzburg 3-Hours race the following day! The circuit was a short twisting one like most of the South African circuits, and the Porsche was really suited to it. The car felt very light and responsive and it was easy to flick it through an ess bend from one lock to another. The basic handling was neutral, converting to oversteer with the power on, and it was a delight to set the car up for a corner and drive through with the rear sliding slightly. However, it did not feel wise to get the rear end too far out as I would imagine it would come right round quite smartly.

The most impressive thing was the performance of the brakes. When I had tried the pedal before starting, it had been spongy, so I had not expected the brakes to work very well. In fact they provided fantastic stopping power with very little effort and pulled up straight with no sign of fade—very comforting at the hairpin at the end of the half-mile downhill straight!

The other satisfying feature was the gearbox which changed very quickly and smoothly; in fact, as quickly as you could move your hand, with no noises or mistakes. As far as the sheer power was concerned, I found this rather disappointing. The car really motors round a circuit, and will do 170 mph on the right gearing, but it just does not feel it is going that fast. Anyone who has heard one of these Porsches will know the fantastic noise they make on a circuit. I was pleased to find out that inside the car the noise was not so penetrating—which was just as well for my eardrums.

Mike de Udy will be racing this car again in 1967, and if last year's performance is anything to go by his name will be featuring prominently in the results. He certainly has a most satisfying car, both for the driver and spectators. Long may it scream!

continental diary

by Paul Frère

Porsche 911R in action at Hockenheim.

More on Bergs and Langhecks

WELL, as Roger Bell challenged me to give my impressions of driving the Porsche 910 at Hockenheim, I gladly take up the gauntlet. But I must jump to his defence for not having been able to describe the happenings when he was taken round the track by Jochen Neerpasch in the very car I drove later on. *My* day started by being taken round in the even more up-to-date eight cylinder engined 907 by Jo Siffert, and I also found that all was over before I could steady myself in a seat that may be shaped to regulations, but certainly does not conform to human anatomy. Somehow, however, I managed to preserve enough lucidity to notice how incredibly quickly Siffert changed down from fourth through third to second for a certain corner—part of the credit being obviously due to the Porsche all-synchromesh gearbox—and, bearing in mind that I was later to drive myself on the Hockenheim short circuit of which I had never seen some of the corners, to note that one of them could be taken, rather incredibly, with the foot nearly flat on the floor. This is a thing I would certainly not have found out myself in the two or three laps I was later allowed to drive.

Having perhaps more opportunity than most of my colleagues to keep in touch with modern racing machinery, I knew what cornering power to expect, and it was rather by contrast that I really woke up to the fact that modern racing cars can go round corners at amazing speeds—the contrast was provided by the ride I took soon afterwards with Vic Elford in the 911R—the racing version of the 911S with about 500 lb less and 50 b.h.p. more. Though this car has much wider than standard rims (6in. at the front and 7in. at the rear) and must handle even better than the normal model which is good by any standard, it felt rather unimpressive, though certainly Elford squeezed the last ounce out of its cornering power. It was admittedly on radial ply tyres, while the suspension is tuned for racing tyres, but this just showed the gap there is to-day between the cornering powers of a first class road car and of a racing car. Some investigation was made by Porsche on this matter, which showed that the maximum centripetal acceleration a 911S could reach on a well surfaced corner was about 0.85g while a 910 would still be manageable at nearly 1.3g. This means that a bend which a 911S just manages at 100 m.p.h. can be taken at 124 m.p.h. by the racing model. At Hockenheim it meant that where Siffert just eased off a little with the 907 sports prototype, Elford had to brake quite hard with the 911R, in spite of the fact that it reached the corner at less speed.

These impressions were confirmed later, when I got a chance to drive the 911R which felt very much what you would expect of any current Porsche, with wide rims, only with more performance on tap (though the gearing was much too high for the short circuit), while the prototype I drove soon after was in an altogether different sphere. This was the 2-litre, six-cylinder engined model 910, which was the mainstay of the Porsche striking power in 1967. The chassis layout differs from the more modern 907 mainly by a slightly narrower (1in.) front track and left hand steering.

I didn't think the performance angle of a car of about 400 b.h.p. per ton which has won both the Targa Florio and the 1,000 km. of Nürburgring outright needs any emphasis, but having "raced" both the 904 and the 906, carrying a television camera and transmitter in the 1,000 km. of Nürburgring for the past two years, I had a good basis of comparison and was anxious to find out how much better the newer car was. With this background, I felt pretty quickly at ease in the 910 and had an idea of what to expect. With lighter and wider wheels and improved tyres, the 910 probably goes round corners even faster than the 906 (or Carrera 6), but only a stopwatch could accurately detect this.

The main improvement lies in the ease of handling. The 906 has a pretty high cornering power but, when it goes, it does it very abruptly and is not an easy car to check. I expected the 910 to behave in the same way, if not even more sharply, due to its even wider Dunlop tyres, but it was just the opposite, possibly due to the fact that the variable toe-in which was a feature of the rear suspension of both the 904 and 906 has been deleted in later models. In fact, the 910 is a very easy car to drive and my impressions were confirmed by the works drivers present who agree it's easier than the previous models. It can be driven with real abandon, powersliding what is really a beautifully balanced and neutral steering car and keeping the tail under complete control by means of the precise and high-geared steering (13:1 around the central position). Driving ease is also provided by the flexibility of the fuel injected engine. Compared with the carburetter version of the six-cylinder engine as fitted to the 906, the Bosch fuel injection system into the induction tracts provides not only 10

Above: 2.2-litre 907 with upswept tail—and very straight-arm driving position. (Left) Jacket off and sleeves up—the author stripping for action. (Right) The 910—a very easy car to drive.

additional horsepower, but also much more progressive throttle response, which greatly helps in balancing the car against the various forces acting upon it in a bend.

Rev limit is 8,000 on the six and the tachometer needed careful watching with the low gearing used, which allowed all four upper gears to be brought into play on the short circuit (first being more of a starting gear), and all gear changes are exceptionally quick and effortless. Though a sort of abbreviated double declutching is normal routine in a racing car, the Porsche synchromesh is so efficient that no more is actually needed than adjusting the engine speed to the next lower gear while changing down, and this must be done pretty quickly, as the lever will select the next gear as quickly as you can move it.

The 910 Berg

Of course, the most interesting of all cars present at the track was the "Mountain" Group 7 car with which Gerhart Mitter has clinched the European Mountain Championship. In general chassis design and dimensions, it is identical with the prototypes, but whereas the latter are built to conform with the minimum weight limit for their class (575 kg. for the 2-litre six and 600 kg. for the 2.2-litre eight), the Bergwagen weighs an amazing 450 kg. (8.8 cwt.) with the heavier eight-cylinder engine. Even bearing in mind that this car is an open model with very scanty lighting equipment, no alternator, no spare wheel and only a 3.5-gallon tank, this is really light. It is the reason why nobody was taken for a ride as a passenger: he would probably have fallen through the paper-thick plastic floor! Neither were any except the works drivers allowed to drive this expensive one-off.

The cost of the Bergwagen has been the subject of much speculation lately, specially by Porsche's main competitors in the mountain championship. Of course special alloys had to be used rather extensively to get down to such a low weight—mainly titanium which, at normal temperatures, has approximately the strength of steel, but weighs only about half as much. In fact this metal amounts to 7.1% of the total weight of the car, including the connecting rods which are made of titanium in all Porsche racing engines (including 911R). Total weight of titanium used in the car is 32 kg. (71 lb.), representing about 2,000 DM (roughly £200), which is nothing in view of the cost of a racing car. Another weight saving factor which has been developed for racing and is just about to be transferred to all production 911 models, is the use of an Elektron crankcase (instead of aluminium) which Ferry Porsche told me was the biggest Elektron casting made in the world.

Speaking of weight and materials, he produced a very interesting breakdown of the various materials used in the mountain car and in the prototypes, based on car weights without fuel, which I reproduce herewith:

	Mountain car	Prototype
	%	%
Light alloy	35.5	27
Steel	32	39
Plastic and rubber	19	23.8
Titanium	7.1	4
Electrical	3.8	3.9
Liquid (Oil)	2.6	2.3

Speaking to the head of Porsche brought to light that much work has been done on the 2.2-litre flat-eight engine and that it is now as reliable as the flat-six 2-litre. It is thus the eight that will be the spearhead of Porsche's racing activities next year, but it is hoped to have at least one full 3-litre engine ready for Le Mans. This, according to Ferry Porsche ". . . will be entirely based upon existing experience", which could well mean that it will be a 12 with the cylinder sizes of the mountain 2-litre eight, though some people in Germany seem to think it will be a slightly enlarged six with two cylinders added to make it a flat-eight. Any of the alternatives would fit a car not much bigger than the existing one, though tyre manufacturers will demand 15in. wheels instead of the 13in. pattern at present used.

911SS?

Speaking of Porsche, rumours are going around that an even faster model than the 911SS—but still part of the 911 series—may well be announced in the near future. Its engine would produce around 180 DIN b.h.p. (20 more than the 911SS) which sounds quite likely in view of the fact that even in its racing form with 210 b.h.p. the 911 engine is entirely reliable. Porsche works driver and specialist Gerhart Mitter already modifies 911S engines to give 180 b.h.p. with, it is said, no loss of flexibility, but quite a useful increase in performance. The German magazine *Auto, Motor and Sport* recently tested one of those cars which, in spite of being a Targa convertible and thus not quite as good aerodynamically as the coupé, reached a maximum speed of 143 m.p.h., did 0 to 100 m.p.h. in 16.6 seconds (3s. less than a normal 911S tested previously) and the standing kilometre in 26.5s., which is clearly in the Ferrari bracket. If Mitter can do this for 2,500 Marks (about £250), there is no reason why Porsche themselves could not do it—and for even less money. **M**

THE TWIST, BUMP AND GRIND OF THE ROCKY AND rolling Targa Florio; the launches and splashdowns of the other-worldly Nürburgring. Car-breakers both, right? Yet this gossamer machine with its translucent bodywork played one like an old fiddle and rocketed into the far distance at the other. Even more remarkably, 10 years later, admittedly in a lesser era of two-seat racing, Porsche's pared-to-the-bone 908/3 was still competitive in privateer hands.

For years Porsche's racing squad had punched above its weight, so much so that people forgot how small the Zuffenhausen concern was as sportscar racing reached its apex. But that only seemed to strengthen the resolve of Ferdinand Piech – Ferry Porsche's nephew – to deliver a knockout blow. When in early 1969, as the tech boss of the Experimental Department, he ordered that the 25 917s required by homologation be rushed through, workers were drafted in from the 911's production line: racing was suddenly the priority, not a side-line. The company was flat out. Make or break. That it was simultaneously working on a car designed purely to win two specialist races is bewildering.

Le Mans was its main aim, and the mighty 917 was *the* car. Targets B and C were the Targa Florio and the Nürburgring 1000Km, for which the mighty 917 was definitely *not* the car. Porsche had a habit of upstaging Sicily's motorsport passion play, but had found it harder to win Germany's second-biggest race: indeed, it had won the Targa seven times before it began its own 'Ring cycle in 1967. It would prevail at the Nordschleife the following year too, in the car that ended its long-accepted role as plucky underdog. For 908 was its first racer based around an engine that sat on a designated capacity limit. Until its 3-litre air-cooled flat-eight appeared, Porsche had always left itself a comfort zone: if it won, everybody marvelled at its underpowered overachievers; if it lost… well, it was expected to. But this softly-softly approach wasn't Piech's bag. He wanted to go toe-to-toe with the big hitters. And by the end of that first 908 season, he was on the ropes. A second Nürburgring Porsche win was the new car's only major success in a troubled year. The marque's famed reliability had deserted it, which was odd because 908 was meant to be the sorted 907 – which won that year at Daytona, Sebring and Targa Florio – with 26 per cent more power.

In truth, the firm had broken its rule of continuous evolution. It had complicated things by building a heavy dry-sumped six-speed gearbox and jettisoning the proven eight-cylinder firing order of the superceded Type 771. These matters were put right in the 908/2 of 1969: in came a simpler, five-speed transaxle, out went the debilitating vibes of alternate power strokes from each bank. The updated 908 was smoother and more usable, and Porsche went for the throat: four, five, even six cars per round, five straight wins (three for the Spyder, two for the *Langheck*), its first world sportscar title – and the nearest of Le Mans misses. Piech was vindicated.

One of the small concessions made by Porsche during this hectic time was the axeing of its European Mountain Championship programme. During the 1960s it had produced increasingly specialised, ultra-lightweight, wheel-at-each-corner cars for this demanding, if scenic, series. The last ➵

PORSCHE'S 908/3 WAS LOTUS-LIKE IN PHILOSOPHY, BUT ITS CRASH DIET DIDN'T STOP IT FROM MUNCHING SOME OF RACING'S ROUGHEST ROADS. PAUL FEARNLEY MEETS WITH A SCANTILY CLAD MODEL PERFORMER…

WHITE

LIGHTENING

But officer, as you can clearly see, it's registered for road use. Top right: Targa testing hack missed out on wild paint job. Below right: Car carried 15 in only contemporary race run

of these was the outlandish 909 Bergspyder of late 1968. The trick to this car was the placing of its gearbox between engine and final drive. This moved the driver forward, centralising mass and sharpening direction-change via improved use of the ever-fattening tyres of the period. It competed only twice, but its legacy would be a long and successful one.

PUSH THE RECESSED BUTTON AND LIFT (WITH ONE finger) the driver's door up and forward. Ahead lies the starkest of cockpits. But what's that? Slowly it dawns on you: that bar running along the bottom edge of the dashboard is the front anti-roll bar. Put it another way: with the 26.5lb epoxy bodywork removed, you can touch the right-front tyre while still strapped to your seat. Put it another way: the rack is mounted above and slightly behind your knees. You're 'cab forward' in a 908/3 because, like the Bergspyder, it's engine-gearbox-diff. Your feet are at least 8in ahead of the axle line and are 'caged' by a spindly frame that carries oil cooler and master cylinders. There is some panelling-in, but this is only thin fibreglass meant to keep small stones out of the cockpit. Let's face it, you're the crumple zone inside this 77lb spaceframe.

"If Colin Chapman had known how much effort went into making the 908/3 light, and how light it was, he would've had a fit," says Vic Elford. "But I felt so secure in it that I never considered it unsafe. It was so controllable that the confidence it gave me was huge." Yeah, but you're Vic Elford; I'm just me. He's right, though, no car has ever felt so alive.

With my precious tootsies offset towards the centre, a typically big Porsche 'tiller' in my gloved mitts and that motor growling in my lug'ole (this is the smooth one?), I nurse the car out onto the… road! Very Targa. Don't worry, we've got a trade plate taped to that peekaboo tail, so it's all above board. And anyway, if I'm stopped by the *polizei* I can explain how road-car it (nearly) all is. For if a driver's safety was not atop Porsche's 'to do' list, his comfort over long distances was: the clutch's feel is long and light, the brakes are activated by a pedal soft enough to make press-on heeling-and-toeing tricky and the gearbox, as ever, is fully synchro-ed. Porsche also melded a go-kart feel with just the right amount of alleviating suspension movement, while the engine pulls hard from the needle-stop. Which

is a good job, because I've pulled away in third!

The gearbox's gate is very narrow and is fitted with an idiot-proof selector mechanism. Sadly, this is proving that I'm an idiot. Worried about dropping into second from fifth, Porsche's engineers made it so that the only option from top (out on a dog-leg for a reason explained later) is fourth. Nor is it possible to hook into second from here: it's third or fifth for you, my lad. That's fine, I've no problem with that. What I do have a problem with is that the gate itself seems canted towards the centre of the car. Instead of an H-pattern, I am faced with what is best described as a narrow W: first is selected by pushing forward and slightly in; second is back and slightly away; third is forward and slightly away, etc. Once again, this is fine – until you miss fourth from fifth. From here, it's back to fifth – or nothing. And once you get into a tizzy, the problem exacerbates. Leo Kinnunen, the Finn who set fastest lap in the 1970 Targa, talked of a stiff gearchange; Björn Waldegård, who made his racing debut with a 908/3 in the same race (see page 92) mentioned being used to nursing Porsche gearboxes because of his 911 rallying experiences. The other surviving 908/3 drivers I spoke to – Elford and Brian Redman – had nothing bad to say about the 'box. Ho hum.

Somehow sensing that there isn't an Elford or a Redman aboard, the engine does its level best for me. Sure, its 911 basis, two-valve head and wide valve-angle was hardly cutting edge at the time, but how much power do you need to hustle a 565kg car? Believe me, 360bhp is plenty. Crack those slides and 908/3 is off like a jackrabbit, nose hunting across the road, sniffing out the next apex. Photographs done, I give it a little squirt. It's not much more than a mile back to our base, but it's a narrow, tree-lined B-road with a couple of downhill sweepers, a big stop at a roundabout and a 90 right. The lady in the Opel gets a shock as this pure-white pure shape crests the rise, but otherwise it's quiet and I get a wriggle on. It's enough to wet the whistle: Targa in a 908/3 must have been mind-bending. This car changes direction like one of those bikes from that failed first 'computer movie', *Tron*. It stops on a *pfennig*, possesses synapse responses – and wows whoever gets in it: Walter Röhrl reckons it's one of the best Porsches he's ever driven....

That one of the rally greats should rave about it comes as no surprise. The motorsport codes find their node at 908/3: all-rounders Elford, Kinnunen and Gérard Larrousse excelled in the car. But then the Targa Florio today is the special stage rally it was, in hindsight, always destined to become. Thirty-three years ago, though, it was still the most challenging of all road races – and Porsche was determined to win it for a 10th time. The car in these photographs – test mule, chassis 003 – was shipped to the island a month before the May 3 event and it pounded round, in and out of everyday traffic, until four brand-new cars arrived for the race.

Porsche's change of attitude was obvious upon their arrival: American Dick Soderberg's beautiful, uncluttered shape was bedecked in the most in-your-face paint jobs yet seen on a Porsche. There was no need to be low-key white any more – they had something to shout about.

Top: Cooling ducts for outboard rear disc brakes.
Bottom: Oil tank for flat-eight's dry-sump system

TECHNICAL SPECIFICATION

ENGINE
Type: air-cooled, twin-spark, DOHC flat-eight
Construction: two-valve aluminium cylinder heads, two-piece mag-alloy crankcase; forged-steel crankshaft; titanium connecting rods; chain-driven camshafts; sodium-cooled valves
Capacity: 2997cc **Bore x stroke:** 85 x 66mm
Power: 360bhp @ 8400rpm **Torque:** 270lb ft @ 6600rpm **Compression:** 10.4:1 **Weight:** 396lb
Carburation: Bosch, mechanical fuel injection
Ignition: Bosch **Firing order:** 1-7-2-8-5-3-6-4

DRIVETRAIN
Gearbox: Porsche, Type 910, five-speed (plus reverse) transaxle, all-synchromesh

CHASSIS
Type: aluminium-alloy tubular spaceframe
Weight: 77lb **Wheelbase:** 2300mm **Track (f/r):** 1542/1564mm **Length:** 3850mm **Kerb weight:** 565kg **Bodywork:** fibreglass **Weight:** 26.5lb
Frontal area: 15.07 sq ft
Cd: 0.46 **Downforce (f/r):** 85/200lb at 155mph
Suspension (f): independent, double-wishbone, coil spring/dampers, anti-roll bar
Suspension (r): independent, single top link, reversed lower wishbone, twin radius arms, coil spring/dampers, anti-roll bar

RUNNING GEAR
Steering: ZF, rack and pinion, unassisted
Brakes (f & r): outboard, vented and cross-drilled cast-iron discs, four-piston calipers **Dampers:** Bilstein **Wheels (f & r):** 15in (13in were used on 1970 Targa Florio) **Tyres:** Goodyear, 240.5x15, (f) 11in, (r) 15in

And how. Kinnunen was almost 1min faster in practice than Elford's best lap of the previous year. And in the race, despite a fantastic effort by Nino Vaccarella and Ignazio Giunti in the unsuited Ferrari 512S, Porsche's latest car scored a 1-2 courtesy of Jo Siffert/Redman and Kinnunen/Pedro Rodriguez.

"When I first saw the car I was worried that it might be difficult to control," admits Redman. "You were so far forward that I wondered if you might not be able to feel what was happening behind you. But it proved to be a fantastic car. Manfred Bantle, the guy responsible for most of it, doesn't get the credit he deserves. He got his sums exactly right: it was nimble, compact and responsive, with fantastic mechanical grip – the perfect combination for the Targa. It had fifth gear off on a dog-leg because you only needed it once per lap on the Targa, perhaps twice on the Nordschleife, and it had lots of low-down torque. For me, it was the first true modern racing car – minus the downforce."

Elford held similar views but met with a very different fate on the car's debut. Vic was reckoned to know every pebble along the Piccolo Madonie's 44.7-mile lap, and was the pre-race favourite. But the rock in the road on lap one was something not even his mental pace-notes could cope with. He swerved, hit a kerb and broke a steering arm that "was about a thick as two pencils". But he and Porsche Salzburg would gain their revenge less than a month later at the 'Ring, where the 908/3 proved 20sec per lap faster than the 908/2 had been in 1969.

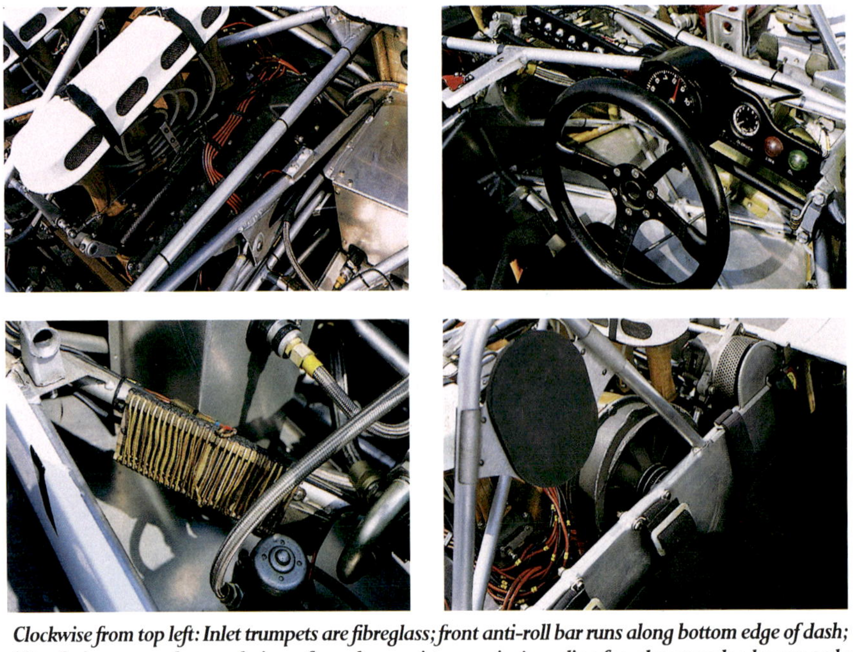

Clockwise from top left: Inlet trumpets are fibreglass; front anti-roll bar runs along bottom edge of dash; bits of wire wrapped around piece of wood – a resistor; engine's cooling fan almost scalped yours truly

There had been fundamental change at Porsche towards the end of the latter year: with the title wrapped up early, Piëch farmed his cars out for the later races. Porsche Salzburg – a Piëch family-owned concern – was the first to enjoy his patronage, and it was joined in 1970 by Slough's JWA. The rivalry was intense: talk to Redman and Salzburg got first dibs on new bits; talk to Elford and JWA was the teacher's pet. At the 'Ring, however, it would appear that it was the Salzburg squad in the know, fitting larger oil tanks to circumvent the problems that sidelined the leading JWA car. Elford, co-driven by Kurt Ahrens Jnr, took the win, while team-mates Richard Attwood/Hans Herrmann were second in 003's only contemporary competitive run.

Porsche would win a second manufacturers' title in 1970 – breaking its Le Mans duck with the 917 – and would continue this form into '71: a second Le Mans victory, a third world title. By this time, however, 908/3's slow-circuit advantage had been whittled away. The 3-litre sportscar formula was looming for '72, and Ferrari and Alfa had joined the party a year early in preparation. Crucially, they had brought F1 engines with them; Porsche had shelved plans for a four-valve 'F1' unit after it had disappointed on the dyno. This meant that 908/3 was ➼

outgunned, although still substantially lighter.

Elford/Larrousse were in contention on the Targa until they suffered two punctures on one lap, thus handing the win to the Alfa T33/3 of Vaccarella/Toine Hezemans. Redman, meanwhile, suffered a fiery first-lap crash: "All they found was a big black hole, at the bottom of which lay the crankshaft!"

At the 'Ring, Elford/Larrousse were able to keep sufficient pressure on the faster Ferrari 312PB and Alfa of Jacky Ickx and Rolf Stommelen to cause them to wilt, and this 1-2-3 result was supposed to be the final hurrah for 908/3. Except it wasn't.

One of the proposed changes for 1972 was the introduction of a 650kg minimum limit. Porsche knew it could not compete on those terms and turned its attention to Can-Am. By 1975, though, as sportscar racing lost its edge, Reinhold Joest, a driver and a team boss with a golden touch, realised that a 908/3 fitted with bodywork copied from the Can-Am 917/10 and boasting a 2.1-litre turbo six from the 911 Carrera RSR racer could be competitive. Three cars were so converted with Porsche's

Top: Car unsuited to banked track – downforce is minimal. Middle: Don't be fooled by its Meccano/Tupperware build. Bottom: Periscope mirror is shape's only protuberance

BJORN TO RACE?

"I STARTED MY CAREER WITH PORSCHE PURELY AS A RALLY MAN. I was signed by Huschke von Hanstein, but not long after that he was replaced by Rico Steinemann, who asked me if I'd like to do some racing. He thought that the Targa Florio would be good for me. I didn't disagree.

"I enjoyed practice because you were allowed more of a free hand in it than you were in the race. Mind you, it was a good job I enjoyed it because, as the newcomer, I always did the first runs, which meant getting up at four in the morning; Jo and Pedro used to turn up at about lunchtime!

"I felt ready for my first lap in a 908/3. It was fantastic and I thought you could be flat everywhere. Then I went off! The big problem I had, though, was that I was too tall; I was crouching so my helmet didn't blow off.

"Those first laps were not funny because it had rained – and snowed! – overnight and there was mud and slush everywhere. But after a long day's work, Richard Attwood and I were fifth. The 908/3 was good for its time, but I'm sure it was set up too hard. You could run it better there now with what we know about suspensions."

tacit approval – and with good reason. As the FIA vacillated between Groups 5 (roost of Porsche's dominant 935) and 6, Zuffenhausen was secretly building a two-seater, the 936, for the latter, and any prior info on its turbo would be useful.

Joest gave them plenty to chew over. Although 908/3 is a spartan machine, it's strong where it needs to be – gearbox casing, engine mounts and hubs appear sturdier than required – and only a little beefing-up was needed to ensure that it could cope with the increased stresses caused by the turbo's extra 200bhp. The gearbox was marginal, but the car was reliable. And quick. Joest scored two seconds, three thirds and a fourth at Le Mans in 1975. In '76, his car won at the 'Ring.

And 908/3 didn't stop there. By 1980, a time when sportscar racing had descended to almost a run-wat-ya-brung level, Siggi Brunn, whose firm was responsible for the rebuild of 003, was running a standard 908/3 in historic *and* contemporary racing when he and Jürgen Barth came within an ace of winning the Silverstone Six Hours. True, the big guns were silent, awaiting the outbreak of Group C in '82, but even when they were loosed off, one of Brunn's now-twin turbocharged warhorses upstaged the new car at the prestigious Norisring race that year, John Paul Jnr leading briefly from pole and finishing sixth with a delaminated rear wing.

The heroic 917 hogs the limelight, but was its little cousin the better car? There is some substance to its argument, for although 908/3 was super-light, it was no lightweight.

Salon

PORSCHE 908/031

From underdog to pit bull

PHOTO BY GEOFFREY GODDARD

BY BRIAN REDMAN
PHOTOS BY BILL WARNER

June 9, 1968, the University of Liège Teaching Hospital, Belgium. Prof. F. Orban, chief of surgery, has volunteered for duty. Today is the day of the Belgian Grand Prix at Spa-Francorchamps, the fastest and most dangerous road-racing track in the world. Experience tells him his services will be required.

Eight laps into the race, I approach the fast left-hand bend, Les Combes, situated at the top of the 5th-gear uphill straight following the pits. Time to brake, but the Cooper-BRM is not slowing. It's better to crash backward than forward, so I try to

turn the wheel to the left...it does not respond...I turn to the right and it turns, taking me closer to the barrier...a last-minute jerk to the left...no response and I crash at over 100 mph.

In dreamlike slow motion, I feel my right forearm break as the Cooper rolls over the barrier. I try to hold my breath as the car bursts into flame; I can't move my right arm; it's hanging in the fire. Fire extinguisher powder everywhere...a course worker, cigarette dangling from his lips, tries to undo the shoulder harness...fire reignites...worker disappears...more powder...coughing and choking...lying on the ground...chaos everywhere. Broken car, broken bodies, broken dreams.

Six months later, having been assured that the bones have healed, I am in South Africa with a 2.0-liter BMW-engine Chevron B8 to contest the six-race Springbok Series. After three races, there is a burning sensation in the forearm whenever it is used.

On December 12, a phone call comes from Germany. "Brian, this is Steinemann, team manager for Firma Porsche. Would you like to drive with us in 1969? The first race is the Daytona 24 Hours on February 1 and 2. You should be in Daytona by the 28th of January."

Would I like to drive?!

December 16, a visit to the leading Johannesburg orthopedic surgeon, David Roux. Twenty X-rays later, the verdict: "Brian, you have no union of the ulna or radius."

"David, I must be at Daytona in six weeks."

"Tomorrow I will operate; don't use the arm until you must."

FOR THE FIRST time in its illustrious history, Porsche is going to compete on a level basis for the World Manufacturers Championship and, perhaps, also for the first time, win the Le Mans 24-hour race. Until now, the Stuttgart-Zuffenhausen manufacturer has always been the underdog competing strongly and with great success with the jewel-like lightweight 2.0- and 2.2-liter cars against the might of Ford with its GT-40 Mk I, II and IV, Ferrari, Chaparral and Lola.

Following the overwhelming wins at Le Mans by Ford in 1966 and 1967, it is, perhaps, not surprising that the racing managers of the European factory teams were hard at work lobbying the all-powerful FIA (Federation Internationale de l'Automobile) to change the rules to exclude the mighty Ford from further participation in the championship and, in particular, Le Mans.

Rumor followed rumor. The redoubtable and efficient Porsche team manager, Hüschke von Hanstein, was in constant contact with the FIA, and although no official announcement came until October regarding limiting engine size to 3.0 liters for the Prototypes in 1968, it is significant that as early as July, 1967, Porsche started work on the design of a full 3.0-liter racing engine. This decision was extremely important, because for the first time in its history, Porsche would attack the championship with a prototype car with an engine equal in size to the competition's.

The rules announced by the FIA for 1968 would restrict the Prototype class to 3.0 liters, the same size as Formula 1. It was hoped this would encourage specialist manufacturer entries, plus Group 4 Sports Cars—considered to be production cars with a minimum of 50 cars produced—to 5.0 liters. The only car in racing trim that fitted these rules was the now somewhat long-in-the-tooth GT-40.

With few cars entered in Group 4 and with Lola lobbying hard for acceptance of its T70 coupe, another late rule change was made by the FIA: The 50-car requirement was reduced to 25. Perhaps a mistake was made in not specifying an engine with a large production basis, for immediately and in great secrecy Porsche embarked on an ambitious and hugely expensive exercise to build 25 5.0-liter *racing* cars.

The story of the 917 will be kept for another day; suffice it to say that the car developed a fearsome reputation.

In 1969, Porsche AG would enter no less than five 908 team cars, with two drivers per car for all the events comprising the World Manufacturers Championship: Daytona 24 hours, Sebring 12 hours, Brands Hatch 6 hours, Monza 1000 kilometers, the tortuous Targa Florio in Sicily, the redoubtable Spa-Francorchamps in Belgium, the difficult 14-mile and 170-corners-to-the-lap Nürburgring, Le Mans 24 hours, Watkins Glen 6 hours and finally the Austrian 1000 km at the Österreichring.

To build, maintain and repair five team cars to run in all these events meant construction of some 15 Porsche 908s and, at the same time, 25 (they all had to be shown to the FIA at one time) 917s. Such a project took every resource that the relatively small auto manufacturer could muster.

Chosen as team manager was the urbane, multilingual (fluent in five languages) Swiss journalist, Gianrico "Rico" Steinemann, an excellent driver in his own right. The engineering staff was under the overall control of Ferdinand Piëch and Ing. Helmuth Bott. Engine development was headed by the experienced team of Hans Mezger, Paul Hensler and Valentin Schäffer. Chassis development came under Helmut Flegl and Peter Falk.

The multinational team of drivers included from Germany the great veteran Hans Herrmann; the very fast and three-time European Hill Climb Champion Gerhard Mitter (soon to perish whilst practicing for the German Grand Prix at the Nürburgring in his Formula 2 car); the competent and quiet garage owner Kurt Ahrens; the extremely fast and brave Rolf Stommelen, the young hillclimb star who was later to lose his life in a Porsche 935 K3 at Riverside; Willi Kauhsen and German aristocrat Karl von Wendt; Austrian Rudi Lins and French rally ace Gerard Larrousse.

From England came the always-smiling and deceptively fast Richard "Dickie" Attwood, the hard-charging and ambitious ex-Monte Carlo Rally winner "Quick Vic" Elford and me.

Competing at Daytona and Sebring was the successful Porsche veteran and American businessman, Joe Buzzetta.

Finally, from Switzerland came the mercurial and fun-loving Joseph "Seppi" Siffert, fated to be asphyxiated under his burning and upside-down BRM GP car at a non-championship F1 race at Brands Hatch in late 1971.

And so, the scene was set for what would surely be Porsche's finest year ever. Full of confidence, the team arrived for that first race of the 1969 season at Daytona Beach. Five beautiful white long-tail *(Lang)* 908/01 coupes with brightly colored nosebands. The main opposition came from four Group 4 Lola T70 Mk IIIB coupes, including a Roger Penske car in Sunoco colors driven by Mark Donohue and Chuck Parsons. Completely rebuilt to Roger's exacting specifications, this car would be the eventual winner ahead of the similar car of Lothar Motschenbacher and Ed Leslie.

Two Ford GT-40 Mk Is were entered by John Wyer Gulf Ford. It is notable that in practice the fastest Gulf Wyer GT-40 driven by the youthful and brilliant Jacky Ickx, the young "Brussels Sprout," had gone from being on pole position in 1968 to 8th on the grid in 1969.

I was, perhaps, the only Porsche driver on the team with any misgivings. My right forearm was weak and most certainly unable to sustain the rigors of 24 hours around the demanding Daytona banking. I ended up driving with my left arm and knee supporting the steering wheel, my right arm resting loosely on top of the wheel in case of an emergency.

It was difficult not to smile when after nine hours of racing, the leading 908 of Attwood/Buzzetta came into the pits with engine trouble diagnosed as failure of the alloy intermediate drive gear from the crankshaft to the camshafts. "Vee are finished; ze vill all break!" They did . . .

Impressions of the long-tail coupe were mixed; it was very fast down the straight but wandered around somewhat on the banking. Behind another 908, you could see the tail out at an angle. Under heavy braking, the car moved about and felt unstable.

Following the Daytona debacle, it was decided to conduct a 24-hour test at Sebring with Seppi and me as the drivers. For this race we would use the short-tail *(Kurz)* open-cockpit car. Torrential rain and Seppi's going upside down at the Hairpin (caused by a cracked chassis) terminated the test.

During the race, the engineers had not calculated on the Sebring track surface breaking up after a few hours, which caused additional stress and strain on the chassis. After four and a half hours, the Herrmann/Ahrens 908 was in the pits and retired with a broken chassis, and shortly thereafter our car succumbed to the same problem. Following the arrival of the third team car with the same difficulty, it was decided to start repairing the chassis with steel bars bolted to the frames. Finally, the Stommelen/Buzzetta Porsche finished in 3rd place behind the Chris Amon/Mario Andretti factory Ferrari 312P and the winning Gulf Wyer GT-40 of Jacky Ickx and Jackie Oliver. It was a portent of things to come.

The furiously fast, aerodynamically precarious Porsche 908 long-tail. Its air-cooled flat-8 engine displaces 2997 cc and makes 350 bhp at 8400 rpm... that is, when both ignition systems are functioning properly.

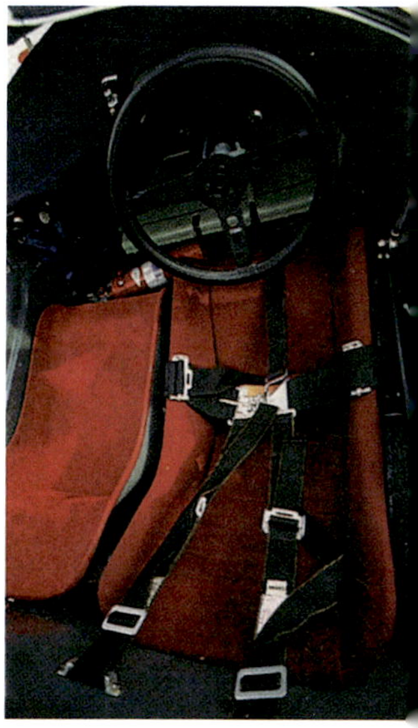

For the 6-hour race at Brands Hatch in England, we again used the 908 open-cockpit short-tail cars. An inspired qualifying run by Seppi put our 908 on the pole. Five minutes to race time, the mechanics made their customary check of the engine's dual ignition system by switching off one system at a time and listening to the engine note drop. Disaster! One system was not working—a loss of 500 rpm and 20 horsepower.

However, we enjoyed an excellent trouble-free race and went on to capture the checkered flag.

Next was Monza and back to the long-tail car. Here on home ground, Ferrari made a big effort and brought two cars, for Pedro Rodriguez and hillclimb champion Peter Schetty, and another for Amon and Andretti. In the race the Ferraris and Porsches battled one another lap after lap. Finally, accidents and mechanical attrition destroyed the Ferrari attack. However, they had shown themselves capable of winning and would be a force to be reckoned with. For the nonce, victory belonged to the 908 coupe I shared with Seppi.

Following Monza came the long trek to that most difficult of races, the Targa Florio. This was a real Porsche specialty race, where prior to 1969 it had already won eight times. Held on 44 miles of normal Sicilian roads, it went through three villages and round no less than 900 corners per lap. Prac-

tice was held on the open road with people, dogs, cats, goats, sheep, donkeys, cars and trucks all competing for space, and in the race you could and should expect anything around the next corner. Unfortunately, Dickie Attwood, my co-driver for this race, visited the Sicilian scenery off the pavement, and we did not finish, but Porsche swept the first four places, led by Gerhard Mitter and Udo Schutz.

Now, back to Spa-Francorchamps. A choice of the new 917 or a regular 908 *Lang*, and we chose the 908. For me a sleepless night before the race, unable to suppress memories of 11 months before, listening to the rain pattering against the windowpanes, listening to moans in the night...were they of love or fear?

The race was a great battle early on, with Siffert in our car and Rodriguez in the Ferrari changing places every lap. On lap 8, right in front of the pits, the battling duo came to lap Karl von Wendt's Porsche 907. Siffert went to one side, Rodriguez the other. Unfortunately, the Ferrari and 907 touched, sending Von Wendt off the track and through a large advertising billboard. The Ferrari suffered body damage, which delayed it in the pits. Seppi and I went on to record another win, and for once I managed the fastest lap.

June 1. The Nürburgring. For this race, Porsche entered no less than six 908s. Three of them had a new smoother body, known as the "Flounder" because it looked like one. In practice at the shorter Südschleife circuit, Seppi managed to crash our race car, landing upside down in a ditch. We were lent a Porsche of Austria regular 908 Spyder, and after another titanic battle with Rodriguez and Amon, we went on to win.

At last: Le Mans. As the earlier-than-normal 2:00 p.m. start approached, the drivers lined up for what was the last of the running Le Mans starts. At the drop of the tricolor, all sprinted across the road, all except Ickx, who made his usual silent protest against this form of start by strolling nonchalantly across the track and making a very leisurely getaway. First away was his John Wyer GT-40 teammate, David Hobbs, swamped within seconds by an armada of white Porsches. At the end of the 1st lap, Rolf Stommelen and Vic Elford went booming by in the 917s, followed by the 908s of Siffert, Schutz and Herrmann, and by the Lola of Jo Bonnier.

Three or four more cars passed, then silence and that most sickeningly all-too-common sight in those days: a thick black plume of smoke rising in the air. At the fast, blind curves just before White House corner, John Woolfe had put two wheels on the grass and lost control of his Porsche 917. Swerving into the earth banking, the car had bounced off and gone back onto the road upside down where it slid before breaking in two and igniting. John, who was thrown out, died on his way to the hospital.

Immediately behind the 917, Amon could not avoid its burning fuel tank; it lodged under the Ferrari's nose, and he was forced to abandon the car and the race. Because of the wreckage across the track, the field came to a halt. When the race continued, the Siffert 908 was able to take the lead. By this time the Stommelen/Ahrens 917 was in trouble with a cracked crankcase, which would lead to its retirement on Sunday morning.

Just before four hours were over, the leading 908 Spyder came into the pits and retired with an overheating gearbox (the car had not been tested with the long tail, and the gearbox overheated because of lack of cooling). Now the Elford/Attwood 917 led with the Ickx/Oliver GT-40 back in 7th place. Two of the 908s were in trouble;

the Lins/Kauhsen car losing 3rd gear, and the Herrmann/Larrousse car stopping for 30 minutes to have a seized stub axle assembly replaced.

After 12 hours, three Porsches still led with the 917 to the fore. After 18 hours, the Elford/Attwood 917 was still in front, though now the engine was sounding flat and leaving enough smoke to obscure the track for following traffic. At 11:00 a.m., the gallant 917 pulled into the pits for the last time. At almost the same time, the 908 of Kauhsen was abandoned at the Mulsanne with no clutch.

The impossible was happening—the young Belgian hero Jacky Ickx, just 24 years old and partnered by Englishman Jackie Oliver, was leading in the underdog GT-40.

The massive Porsche onslaught had dwindled to one car, the 908 of Herrmann and Larrousse, in 2nd place on the same lap. With an hour's racing left, the two cars were racing side by side. Ickx was told to take the engine to 7000 rpm (6000 had been the limit until now), while the Porsche, the faster car, was somewhat off-song. When the dueling duo came up to lap Mike Hailwood in the second Wyer car, somehow Herrmann had great difficulty in getting past (!) and Ickx was able to gain a few precious seconds. The combatants passed and repassed. Now the last lap: First Herrmann was in front, then Ickx; the low-brake-pad light glowed its warning on the dash of the 908, and Herrmann was forced to brake early for the Mulsanne corner. Ickx slipped through.

Once past, there was no way for Herrmann to retake the lead, and he finished a tired and disappointed 2nd just 20 meters behind the winning GT-40; it was the closest "real" finish in Le Mans history.

After Le Mans, Porsche would go on to win the two remaining rounds and finally clinch the coveted world championship.

As with many old racing cars, Porsche 908/031 went through several changes. The coupe body was removed and a Flounder body fitted before being sold to the Aase brothers, West Coast race drivers and Porsche parts specialists. It was to be a backup to their regular two-car Can-Am team, but was never used before being sold next to John Bond Jr., the son of former *Road & Track* publishers John and Elaine Bond. The present owner, Floridian Gerry Sutterfield, a collector and restorer, bought 031 in 1982 and continued to race it in East Coast vintage events. In 1988 a total restoration to 1969 Le Mans configuration was started, great attention being paid to authenticity with much help given by the factory.

I recently had the privilege of driving her at a rather damp Roebling Road race track near Savannah, Georgia.

Carefully opening the door and sliding into the comfortable upholstered seat, I feel the years slide by. Surely it's not 24 years? More like four! First, I look around the cockpit; surely it was never this pristine? No, it wasn't. Built with ultimate lightness in mind, the original cars were almost translucent, whereas this car looks ready for a show.

I start the engine with the usual Porsche ignition key, and she fires immediately, settling down to a steady throb whilst completing the warmup. Soon the oil temperature is high enough, 1st gear is engaged and out we go onto the track. Lapping steadily for several laps to allow the transmission and brakes to warm before feeling again the powerful thrust of the engine, the superb brakes and sensitive steering.

It is not hard to imagine one's self back at Daytona, Sebring, Monza, Spa, Le Mans...not hard to imagine one's friends, seldom seen, many gone.

Special K

Following in the wake of the highly acclaimed 550 series, the 718RSK promised to be one of the finest Porsche sports-racing cars of all time – as, indeed, it became. Keith Seume looks back over the history of the 718 and its legendary four-cam engine. Photography by the author

One tends to get used to the idea of everything going right for Zuffenhausen, and particularly after the impressive debut of the 550 in the mid-1950s it would have been easy to assume that every subsequent Porsche would automatically be a similar success. It wasn't to be, however, as the factory would find out the hard way when the 550's successor eventually appeared a few years later.

But that's getting ahead of ourselves; let's see where it all started. The 550 first saw the light of day in 1954, powered by the 1498cc Type 547 engine with twin overhead camshafts per bank of cylinders. It was a direct descendant of the original Glöckler Porsches assembled by Walter Glöckler, a Volkswagen dealer in Frankfurt, who had taken it upon himself to build a mid-engined Porsche-powered sports car as early as the winter of 1949/50.

With a Type 369 engine of 1086cc producing a modest 58bhp on methanol fuel, the Glöckler Porsche was an immediate success, allowing its owner to capture the German championship in his very first year of competition. A second version soon followed, this time using the 1.5-litre Type 502 engine producing some 98bhp (again on methanol), and again Glöckler captured a German championship title.

Despite this and a number of other successful forays into world-class motorsport, however, it was becoming increasingly obvious to would-be competitors that, in order to stay on top – or anywhere near it – any effort relying on Porsche-derived race cars would need to enjoy the full support of the factory. But at least the foundations had been laid.

Work on a successor to the Glöckler cars began at Zuffenhausen in late 1952. The new machine was another mid-engined design, powered once again by a relatively standard 1500S engine of 1498cc running on methanol-based fuel. One of the first major outings for the car, known as the Type 550, was at Le Mans for the famous 24-hour race in 1954. Regulations dictated that ordinary pump petrol be used instead of methanol, and this resulted in a power output of only 75bhp instead of the planned 98bhp.

Not a single factory-entered RSK completed the race, each one succumbing to engine failure which was later traced back to that roller-bearing crankshaft

In order to combat this the racing 550s were equipped with a streamlined teardrop-style aluminium hard-top in an effort to increase straight-line speed. It worked, too, for the two team cars, driven by Helm Glöckler (Walter's cousin) and Hans Hermann, finished in first and second places in the highly competitive 1500cc class. The 550, Porsche's first purpose-built race car, was well and truly on the map at last.

The production version of the car, the 550-1500RS Spyder (RS was short for Rennsport, German for motorsport), was powered by an all-new four-cam engine of 1498cc (see panel on page 56) and continued to be campaigned with considerable success by the factory drivers. And the simple tubular chassis, with torsion-bar suspension at all four corners, won much favourable comment for its rigidity and fine handling characteristics.

By 1954 the first customer cars were made available, these being snapped up by several teams, most notably in the USA. Under the watchful eye of team supremo, Huschke von Hanstein, Porsche continued to use the 550 as the backbone of its racing effort until 1956, when a revised version, the 550A, was built to contest the Mille Miglia event in the highly capable hands of Hans Hermann.

Powered by a 130bhp version of the four-cam motor, this new variant benefited from a revised chassis design which

Developed from 550A Spyder, 718 (main pic and below) was unveiled in 1957. Although rather slab-sided, it was undeniably aerodynamic. Fuel tank (right) mounted at front

Special K

featured a full spaceframe layout (the earlier model had a simpler tube chassis, remember) and redesigned rear suspension. Because of a lack of time before the event only one 550A with the all-new chassis could be built, and the remaining factory cars retained the earlier chassis.

The 550A was undeniably a superb car, but Zuffenhausen's engineers were already embarking on yet another new project. Hence the 1956 appearance at the Solitude GP circuit near Stuttgart of a shorter, narrower version of the 550A. It was powered by the usual quad-cam engine, but the suspension, although still based around a swing-axle design and featuring Porsche's beloved torsion bars, was significantly revised in detail.

It wasn't a success, however. Thanks largely to what was said to be its nervous handling in a straight line it was quickly nicknamed Mickey Mouse by some drivers, and only an accident at Berlin's high-speed Avus circuit, followed by a catastrophic fire which totally destroyed the car, saved the factory from prolonged embarrassment.

Into 1957 and Porsche silenced any critics who might have suggested the company had been dragging its feet in the search for a suitable replacement for the 550 when it unveiled a new car literally minutes before the start of practice for the Nürburgring 1000km event. This machine, a substantially redesigned 550A Spyder, was much lower than its predecessor, more aerodynamic, and featured a small vertical stabiliser fin on each rear quarter.

The front suspension was strengthened by a special 'K'-shaped crossmember, and it was this that earned the vehicle the designation 1500RSK; officially it was known by its project number, Type 718. Interestingly enough, in later years, even when the 'K'-shaped member was dropped from the chassis, the model still retained its RSK designation. Traditionalists were in for something of a shock at the rear, too, with the replacement of the earlier torsion-bar springs by a pair of so-called coil-over dampers.

The 718RSK, as it inevitably became popularly known, was made available to customers in 1958, powered by a 1498cc version of the four-cam motor developing a maximum of 142bhp at 7500rpm. The following year the power output would be increased to 148bhp at 8000rpm. And the RSK promised to be a competition success from the very beginning, quickly

The heart of the matter

SO, WHAT MADE THE 718RSK tick? The heart of the design was the well-proven air-cooled Type 547 engine, first put to good effect in the 550. The 1498cc unit, based around a Hirth roller-bearing crankshaft, had a bore and stroke of 85mm and 66mm, respectively, the latter seemingly a throwback to the old days when Porsche relied on VW-derived powerplants.

The first examples used in the 718 models produced 142bhp at 7200rpm, but a change of camshaft profiles and other minor details soon pushed this up by 6bhp. For most situations, incidentally, factory drivers were limited to the aforementioned 7200rpm, although they were *in extremis* allowed to use up to 8000rpm for short periods. Peak torque was a modest 108lb ft at a relatively low 5900rpm, while the compression ratio was set at 9.8:1 (although it was occasionally reduced to 9.4:1 according to the event).

What made this engine so special, however, was the complex but almost fiendishly clever means by which the double overhead camshafts per cylinder bank are driven. Via a pair of reduction gears, the crankshaft drives a separate auxiliary shaft at half engine speed. From this auxiliary shaft a set of bevel gears drives two horizontal drive shafts, one for each cylinder head. These shafts in turn drive the lower (exhaust) camshaft in each head by way of another pair of bevel gears, the drive to the upper (inlet) camshaft being taken by yet another pair of shafts to yet another set of bevel gears...

Altogether, there are no fewer than six sets of bevel gears whose backlash (mesh, basically) has to be very precisely adjusted, and it is also absolutely vital to set the end-float on each of the four camshafts correctly. Oh, and then you still need to set the valve clearances on the finger-style rocker arms, of course. Small wonder, then, that four-cam Porsche engines were – and remain – among the most expensive to maintain.

As far as the induction system was concerned, the four-cam motors relied on a pair of Solex 40P11 dual-throat downdraught carburettors for the 550 model, but these were subsequently replaced by Weber 40DCMs in the first of the 718s. These were later changed to 46IDMs.

All the four-cam units featured two spark plugs per cylinder, but the first examples also featured a pair of distributors, each driven from the end of the relevant inlet camshaft. This design proved to be unreliable, however, and it was superseded by a system which placed the two distributors in vee formation and driven by another set of bevel gears from the end of the crankshaft. (That makes a total of seven!) Lubrication was by means of a conventional dry-sump system, the pump being driven by the same auxiliary shaft used to drive the camshafts.

Undoubtedly one of the most complex engines of its time – and it still had only two valves per cylinder, remember – the four-cam Type 547 motor none the less remains one of Porsche's best-ever designs. With a characteristic rattle from the roller-bearing crankshaft, accompanied by the unmistakable howl of the high-revving cooling fan, the engine sounded like no other. But then all Porsche engines sound just that little bit special, don't they? ■

Complex flat-four (left) featured four shaft-driven overhead camshafts; note downdraught Webers

Cockpit is cramped; windscreen vestigial. And Moss, Behra et al raced without benefit of roll-over protection later fitted to this car...

notching up a third place behind a pair of works 3.0-litre Ferraris in the 1958 Buenos Aires 1000km race. Drivers for this auspicious occasion were Stirling Moss and Jean Behra.

The car achieved a similar result at Sebring, while Jean Behra and Jürgen Barth piloted it to a commendable second place in the same year's Targa Florio. But Le Mans was the main target of the racing year, of course, and no fewer than five 718RSKs were entered for the event. Although one car was eliminated in an accident, the remaining four finished in third, fourth, fifth and ninth positions, and only a pair of rival 3.0-litre cars (a Ferrari and an Aston Martin) prevented a complete Porsche whitewash of the event.

An interesting variation on the theme was unveiled in 1959 when the 718/2 made its appearance. This was a single-seater based on the standard 718RSK, complete with all-enveloping bodywork and a simple aluminium tonneau cover to fair in the cockpit. The purpose of this hybrid? To put a toe in the water of the very competitive Formula 2 championship ahead of the company designing a purpose-built single-seater.

Lucky 13

THE 718 SHOWN HERE, CHASSIS number 718-013, was photographed in southern California, where it was being offered for sale at the time for a sobering $425,000.

Subsequently shipped to a fortunate new owner in Japan, the car was originally campaigned by Otto Zipper, Ken Miles and Jack McAfee, three names synonymous with Porsche racing in the USA in the late 1950s. The subject of a recent flawless cosmetic and mechanical restoration, it is one of the finest survivors of an era when Zuffenhausen's products totally dominated world sports-car racing.

Aside from the now mandatory and unfortunately somewhat ungainly roll-over bar, this is as factory-fresh a 718 as you are likely to find anywhere. Sleek, silver and straining at the leash, it simply begs to be raced. One can only hope that it won't be forced to see out its retirement in air-conditioned comfort. ■

Engine bay was vented by means of these slotted panels, removable for access

News of a change in Formula 1 regulations for the 1961 season, which would allow engines of no larger than 1500cc, had proved to be too great a temptation for Porsche, hence its apparently sudden interest in single-seat racing. And despite its sports-car origins the 718/2 proved to be a strong contender in this category, too, Jürgen Barth taking a worthy second place at the Nürburgring, with Masten Gregory taking a class win at Avus soon after.

The 718RSK continued to be the staple diet of Porsche racers throughout the 1958 and 1959 seasons, but success was not as readily achieved as Porsche had become accustomed used to. At Le Mans in 1959, for example, Ferry Porsche took the decision to enter all the team cars with revised engines featuring higher-lift camshafts and a new Hirth crankshaft with larger bearing journals.

It proved to be a serious error of judgement, however, for the once-reliable four-cam engines were stressed beyond their limits during this marathon event. Not a single factory-entered RSK completed the race, each one succumbing to engine failure that was traced back to that roller-bearing crankshaft.

Worse was to follow at Avus, just over a month later. The team car, driven by the very popular Jean Behra, went out of control on part of the high-speed banking and struck an old gun emplacement at the side of the track. Behra was thrown out of the car and killed instantly. Wolfgang 'Taffy' von Trips went on to win the event for Porsche, but it was a hollow victory.

> *The 550's simple tubular chassis, with torsion-bar suspension at all four corners, won...favourable comment for its rigidity and fine handling*

By now, the basic design was some six years old, the 550 and 718s being very much from the same mould, despite their differences in suspension design. But the RSK continued to be campaigned to great effect in the hands of privateers, particularly in the USA where SCCA (Sports Car Club of America) events often appeared to be completely dominated by the German machines.

In 1960 Porsche was forced to rethink the 718RSK, however, when new regulations decreed that sports cars should bear a closer resemblance to 'real' cars (sound familiar?). The result was the RS60, essentially a 718RSK with an extra four inches (10cm) in the wheelbase, and a wider body. Not surprisingly, the changes found favour with most drivers, who considered it to be a better-handling car than its immediate predecessor. And at just $9000 apiece, the 14 RS60 customer cars soon found good homes.

A year later came the launch of the essentially similar RS61, certain versions of which were to be powered by an all-new eight-cylinder Type 771 engine with a capacity of 2.0 litres. By now the writing was on the wall for the 550/718 series, however, as Porsche had somewhat grander plans for the future, involving six-cylinder engines and, of course, chassis numbers commencing with the figures nine and zero... The 1950s had been truly great years for Zuffenhausen, though, and the way ahead looked equally bright. ■

Engine	Air-cooled flat-four, mid-mounted, Type 547/3. Light alloy crankcase and cylinder heads; cast-iron cylinder barrels. Four main bearings; two shaft-driven overhead camshafts per cylinder bank operating two valves per cylinder; two Weber 46IDM downdraught carburettors
Capacity	1498cc
Bore/stroke	85.0mm/66.0mm
Compression ratio	9.8:1
Maximum power	148bhp at 8000rpm
Maximum torque	108lb ft at 6300rpm
Output per litre	98.7bhp/72lb ft
Transmission	Cable-operated single-plate clutch; five-speed rear-mounted transaxle with synchromesh on upper four forward gears; ratios changed according to circuit and type of event
Suspension	Front: independent with multi-leaf torsion-bar springs; spherical joints throughout; hydraulic dampers. Rear: independent via swing axles and Watts linkage; combined coil springs and telescopic hydraulic dampers
Brakes	Hydraulically operated 280mm aluminium drums with cast-iron linings front and rear; dual-circuit operation
Wheels	Steel rims with cast-aluminium centres, or all steel; 4J x 15 front and rear
Tyres	Front: 550x15R Rear: 590x15R or 600x15R
DIMENSIONS	
Length	3600mm (141.7in)
Wheelbase	2100mm (82.6in)
Track	Front: 1290mm (50.7in) Rear: 1250mm (49.2in)
Weight	530kg (1168lb)
PERFORMANCE	
Maximum speed	140mph (approx)

RSKs all round

OF ALL THE PORSCHE REPLICAS that have been produced over the years, the GP Spyder is probably by far the most successful – both visually and commercially.

Launched in 1983 by former beach-buggy manufacturer GP Vehicles Ltd of Isleworth, Middlesex, it was was available from the start either as a kit for home assembly or as a complete so-called turn-key vehicle.

Mechanically the GP Spyder has always owed rather more to Wolfsburg than to Stuttgart – it sits on a specially shortened VW Beetle floorpan, and can be powered by virtually any of the company's air-cooled flat-fours – but that doesn't matter a bit.

Thanks to a wide range of options such as replica Porsche instruments and wheels – and even an RS60-style full-width windscreen – it is possible to create a convincing RSK lookalike for a fraction of the cost of the real thing: kit prices start at just £2100 plus VAT, complete cars at around £13,000.

Contact GP Projects, Unit 49B, Princes Estate, Princes Risborough, Bucks HP27 9PX; tel: 01844 275202; fax: 01844 275230.

Beetle-based GP Spyder is a convincingly close replica of classic Porsche; today far more practical, too (Photo: James Mann)

continental diary

by Paul Frère

Porsche: how to win races

PORSCHES were outstandingly successful in racing last year and even more so from the beginning of the current season, when they scored overwhelming victories both in the Daytona 24 Hours and Sebring 12 Hours races and were only very narrowly beaten in the Brands Hatch 500 Miles. My main purpose when visiting the two leaders of the racing and development team, Ing. Piëch (Ferry Porsche's 29-year-old nephew) and Bott, was to discover what policy had brought such success. It is true that the new regulations limiting engine capacity to 3 litres for prototypes and to 5 litres for so-called sports cars did help put Porsches even more in the picture but, up to now, their cars are still a lot smaller engined than their competitors', in spite of which they have not only the speed, but also the reliability which many rivals are still vainly seeking.

Though Porsches have come very much into the limelight and apparently go racing in a much bigger way than heretofore, it was emphasized that the racing budget in the factory had not grown in any significant way for many years. Only the policy has changed: money is spent building many brand new racing cars, as required by the racing programme, but a considerable amount is recuperated at the end of the season, when the cars are sold to private owners who, in the case of last year's successful 910 models, willingly paid £8,000 to £9,000 apiece. Altogether 25 racing cars were built last year and 18 were sold, the rest having been dismantled or kept for various development tests.

The idea of building so many new cars is to avoid using the same car for two long races run over tough circuits likely to fatigue the structure. This policy was decided upon when trouble was experienced with breaking chassis tubes in the Nürburgring 1000 kilometre race of 1966, on cars which had previously run in the Targa Florio. This year, even more racing cars will be built than last year, and I could see at least 30 in various stages of completion—some no more than a chassis frame—in the racing shop and in stores outside. Their ultimate aim may be to build enough cars to get them homologated as sports cars by the time they are sold. Such a policy implies real confidence, since customers are not likely to be numerous or willing to pay high prices for cars that don't win races!

This is but one example of the logical thinking and acting which has put the Porsche Racing Department where it is now: whenever something undesirable happens, the reasons for the mishap are thoroughly investigated and measures are taken to avoid a recurrence. For example, the current 907s have front hinged doors so they can't fly open even if they are not properly shut on the occasion of a hurried start—no more "gull-wings"—and the rear bonnet is also hinged at the front on all cars designed since Mitter had his accident at Spa, when the rear lid flew open and more or less lifted the car off the road. Both solutions slightly increase the weight, but enough can be saved elsewhere to bring the total down to the minimum weight limit for prototypes.

Porsches have become really fussy about aerodynamics. Not only have they developed long and short bodies (with appropriately extended chassis frames to take the long tail version, the short versions to be used only on short circuits where a long tail would be too cumbersome) but their aim is to reduce the number and size of forward-facing air intakes to a minimum, as wind tunnel tests have shown even very small inlets to increase the drag quite considerably; they are very confident that the 12 h.p. required by the cooling fan of the air-cooled eight-cylinder 2.2-litre engine

At Nürburgring—Porsche driver Jo Siffert in conversation with Huschke von Hanstein of the Porsche organization.

at peak revs is far less than the power required to push a forward-facing radiator intake through the air at high speed. (On slow circuits, of course, the advantage is less and may even become negative.) Another point which is often overlooked is that an air-cooled engine requires less air flow than a water-cooled one, as the heat transfer rate is much higher from a cylinder wall and head, at a temperature of around 300° C, than it is from a liquid radiator at a maximum of 120° C.

Even the modern roller skate racing tyres are looked at askance because of their width which, in the end, results in a wider body, a larger frontal area and increased drag and even the latest cars use comparatively narrow section rims and tyres because it is felt that on fast courses, with considerably less power than many competitors, a reduction of the aerodynamic drag pays off more than a few additional square inches of rubber on the surface. For slow courses, slightly wider wheels and tyres are used in conjunction with the short-tail body.

The races where the comparatively small-engined Porsches have the best chance to beat the opposition are the Targa Florio and the 1,000 Kilometres of Nürburgring, which both call for soft springing. This in turn calls for a suspension geometry with considerable camber change, to keep the outer wheels at least perpendicular to the ground when the car is cornering, and a suspension with a lot of camber variation can only be accommodated by tyres with a sufficiently rounded tread surface. These are the factors summing up to the continued and successful use of tyres which are presently considered obsolete by Dunlop. As a matter of experiment, however, we may see one or two of the team cars fitted with low profile 15-inch tyres running at Le Mans.

Engine endurance testing has been made much more severe, following an analysis of how long and how often the engines actually ran (a) at full load; (b) at full load *and* speed on the Le Mans circuit, and the main single contribution towards making the present eight-cylinder engine thoroughly reliable was a new crankshaft design doing away with the Hirth serrations to secure the flywheel. By the way, this engine uses sodium-cooled valves both on the inlet and exhaust sides. How utterly reliable this grown-up, ex-Formula 1 unit has now become was proved by the fact that the winning car in the Daytona 24 Hours Race had previously done a complete, full length rehearsal of the race, plus all the practice both for the rehearsal and for the race, without the engine being touched—not even for a check on the valve clearance—and had thus done about 50 hours' racing when the car crossed the finishing line.

Another principle of the house is that the last few horse-power are very hard to obtain and to make reliable, and that it is thus much safer and cheaper to have a few less and design a car that needs less. This is why

Racing to improve the breed

At Porsche, the racing department and experimental department for future production models are one single unit, run by the same people, and racing is considered an important part of the general development work. One of the most interesting racing developments that has already been passed on to current production 911s is the die-cast magnesium crankcase which weighs 28 lb. instead of 53 lb. for the former aluminium alloy part. Further weight—and price—savings have been made by replacing sheet metal ducts and cowlings, of which there are quite a few in an air-cooled unit, by moulded plastic (which has also made the engine quieter) and altogether some 44 lb. has been saved on the 911 engine alone since it went into production in 1962. This is important not only in itself, but especially because with an engine overhung at the rear the weight distribution is considerably improved.

It can even be said that the racing car development is being systematically oriented

Scene of activity at the Sebring pits.

Porsche: how to win races

towards solutions directly applicable to production cars. An example of this is the car specially developed for the European Hill Climb Championship, which has been brought down to the remarkable weight (for a 2-litre two-seater car with an all-enveloping body and lighting equipment) of 990 lb. by the extensive use of titanium among other things. This was a costly but time-saving way to get weight down without having to do any major redesigning or development work. Now the second phase is under way, consisting of replacing wherever possible the titanium with light alloy at no cost of lightness, in order to pass on the weight saving to production cars at an economical price.

Future production is also one of the reasons why the new 3-litre engine is being developed to replace the current 2.2-litre unit in the racing cars. As is generally known, the latter is a development of the 1962 Formula 1 engine which had been designed purely as a racing unit and could never be anything else: Piëch told me that whereas the six-cylinder 2-litre engine used up to last year, which was practically a tuned version of the 911 with the injection system added, could be assembled in a matter of 14 to 16 man-hours, it takes 220 hours to assemble the 2.2-litre eight-cylinder! The new 3-litre, which is also an eight-cylinder, has been designed with possible production very much in mind and has a simple chain drive to the four camshafts and a belt-driven fan. Design work on it began in July of last year and the first engine was put together four months later. Development took another four months and those responsible for it are confident that it has now acquired enough reliability for a 1,000-Kilometre race and they can only hope that it will be capable of lasting 24 hours by the time of Le Mans. This is pretty quick work made possible only by the fact that the new 3-litre is a direct extrapolation from the experimental 2-litre six-cylinder with four camshafts made last year which started from the basic crankcase dimensions of the 911 but was used only in a practice car that went to the Circuito del Mugello.

Piëch and Bott make it quite clear that Porsche have a 3-litre car ready early in the season only because the 3-litre limit for prototypes was anticipated and the factory wanted to build cars to compete for top honours, but they would never have been ready if they had waited until the CSI finally adopted the limit; by this time many parts for the new engine had already been made. One of the main problems is the delivery time asked by many suppliers: for example, it takes four months to get a crankshaft, which presumably is the reason why the 2.2-litre, which proved to be unreliable last year, could not be made to surmount its weaknesses before the season ended. Perhaps such plain and obvious facts will induce the CSI to think twice before they again introduce new rules at four months' notice. Porsche themselves would possibly have chosen to make a 5-litre, had they known in time that this limit would be adopted for homologated sports cars, as with their present policy of making a great number of racing cars and later selling them, it would not have been difficult for them to make enough to get them homologated.

As a possible future production unit, a 3-litre looks a better proposition, however, and in spite of having been designed with this in view, it is only 22 lb. heavier than the 2.2 and very little longer: it fits easily into the existing wheelbase and only the bulkhead on the passenger side had to be brought forward a little to clear the foremost of the two staggered banks of cylinders. With a bore and stroke of 84 x 66 mm., it develops a maximum torque of 220 lb./ft. at 6,400 r.p.m. and the power is 310 h.p. at 8,000 r.p.m. This is an official and—I was told—true figure, but nobody said it was the *maximum* power, which it is obviously not, as the engine will rev considerably higher. There being no room for a multiple disc clutch, as required by the torque output, in the normal position between engine and gearbox, the clutch was moved to an overhung position behind the gearbox, where a three-plate unit is used.

Compared with the smaller engined cars, the weight distribution of the 3-litre is influenced less by the engine than by the new all-synchromesh six-speed gearbox. This has dry sump lubrication with a series of small jets, accessible from the outside of the casing into which they are threaded, which squirt lubricant on to the various gears. It had been designed before the decision was taken to build the 3-litre car for use in the hillclimb model and with some margin for future torque increases, but when it was found that it weighed 55 lb. more than the existing five-speed box, it was promptly shelved again. For the 3-litre however, this is of little moment, as the minimum weight required by the rules is 50 kg. (110 lb.) more than for the 2.2-litre, and with the tanks moved slightly forward to compensate partly for the increased weight on the rear axle, the all-up weight distribution is still of the order of 40/60.

Apart from the compulsory weight increase, the chassis and body dimensions of the 3-litre are still exactly the same as for last year's 2-litre, six-cylinder models which developed 220 h.p. and already reached 186 m.p.h. down the Le Mans straight. If Porsches can really get away with their present wheel, tyre and brake sizes, so that the present dimensions can be kept, the 3-litre, with nearly 50% more power available, should be really worth watching. **M**

The 908 spyder, looking like a cross between an airport tow truck and a well-used belly tank, stands hip high to a 911 R six with those four cams.

RACING PORSCHES--'69 STYLE

Strategy is to concentrate on the Manufacturer's Cup and the builder's half of the rally championship, leaving hill climbing to private owners.

by Sloniger

Look out world, here comes the 1969 Porsche racing team with twelve, count 'em, twelve hot pilots, a new team boss and a fresh spyder, more ponies and a tauter competition schedule. Zuffenhausen means to win a few.

Strictly speaking the white cars with suspension-operated flippers mean to win fewer titles in '69 but bigger ones. They are giving the European hillclimb title a bye for the first time since 1963, "to give our private drivers a chance," and skipping the driver half of the rally wreath.

Porsche will contest all events counting towards the '69 Manufacturer's Cup (the one they lost by a whisker to JW Fords last year), and the builder's half of the rally championship. Once again this year rally drivers can run for either personal points or company glory but no one rally counts for both. Porsche-contract hardies will, repeat *will,* keep their eyes on the marque title.

One exception is the Monte where they have Elford trying to repeat, Toivonen the defending rally master, Waldegarrd who Swedes reckon to be their quickest boy by far and Larrousse the ex-Alpine pilot who came so close to copping the Monte in 1968. If you can't beat 'em, hire them for your own team. That's four absolute front-rank drivers, just a third of the Porsche pack, and Elford is only rallying on the side.

Vic will also run big sports prototype events for Porsche and in the best of company. Along with Jo Siffert, one of the more under-rated GP pilots, they have Mitter and Herrmann back too, plus young Stommelen who needs a firm hand. (He shunted a spyder during warm-up laps for their traditional January press do.)

Porsche's chief problem this past season was named Ford, although Alfas always looked to stealing a jug of wine or loaf of bread if anybody got careless. Ergo, two of the better Ford pilots, names of Attwood and Redman from England, are now Porsche chauffeurs. Just to make sure Alfa doesn't suffer ressurance pangs either, Porsche grabbed back Schutz too, a one-time Porschist who rusticated a year in red machinery. Finally, they added Germany's only ranking F2 driver, a blocky item named Ahrens. This is quite some dozen to keep the opposition honest at sites like Sebring, Le Mans and of course the Nurburg Ring where white cars better win or *giffs donnerwetter.*

Talk about a new broom! Even the team boss cum publicity panjandrum checks in fresh this year. The oft-praised Huschke von Hanstein was retired upstairs in favor of a Swiss of many languages named Steinemann. He's the same 30-year-old ex-editor named Rico who at least buttered Porsche bacon with a private Le Mans second in '68 after the works cars dropped out. Henceforth he must find ways to prevent a replay of that event.

As weapons for this assault on Ford foundations, Porsche offers three basic track varieties plus a couple of GT and/or Gp 5 flavors with not a vanilla in the lot. Prime cannon in Zuffenhausen's armory is a 908 available to their legions in closed or spyder form, depending on course. The new open one didn't appear at Daytona but did at Sebring. And where a track is really extended there is a 908 long-tail too. In short, they have the three-liter closed pair which wrapped up '68 plus a lighter and slightly more compact open one for twisty tracks.

Remember, windshield height is a matter of conscience now and rules have eased generally, making the return of spyders a certainty. At least, they'll be on tracks where a performance weight of 4 lbs/hp but "only" 175 mph top are more vital than the ability to touch 200 mph while weighing 4.3 lbs/hp. Such are the differences between spyder and long-tail, with the bobbed but roofed 908 falling between.

All use their flat-eight of three liters and 350 proper ponies. Small wonder the four-cam, light-metal engine gets a 908 there first so often. Incidentally the only admitted difference between spyder and coupe is a single-plate dry

From left to right we have three-quarters of Porsche's '69 racing might: Hans Herrmann, Bjorn Waldegaard, Pauli Toivonen, Rolf Stommelen (making snowballs), Kurt Ahrens, Vdo Schutz, Brian Redman, Dick Atwood, Gerard Larrousse and, of course, the 908 spyder.

The 907, or how to terrify journalists in a snowstorm and collect a basket of fouled spark plugs.

clutch for the former and three plates for closed 908's.

A great deal of titanium goes into such items as fly-wheels, con rods and even fuel tanks when you get to the hill car which Porsche won't do these days anyway. They'll leave these to Mitter.

That 950 lb. light-weight on an 89-inch wheelbase, so short that a driver's feet had to precede the axle if you wanted to get him all in, might not do more than 155 mph but since it only boasted a meagre 4 gallons of fuel, the point remained academic. The juice was carried in a titanium ball weighing a mere 26 ounces, and by pressurizing this globe Porsche also eliminated fuel pumps.

All of which, and beryllium brake disks to boot, explains a 3.44 lb/hp figure for the stubby mountaineer. Despite a capacity limit of two liters,

Porsche extracted 275 hp from their uphill sprinter or 137.5 hp/liter, the best of their range quite naturally. It is a shame we can't possibly see more direct carry-over from this 909 for at least another season, given average Porsche rumination.

Another of the old guard on their phase-out list, though it brought in several '68 victories (Targa for one) is the 2.2 liter, eight-cylinder 907, a more highly-stressed machine than even the 3-liter cars and no heavyweight at 1325 lbs. for its 270 hp.

The factory saw fit to entrust this beast to a few journalists with more eagerness than sense. You see, it had begun to snow big oily flakes and there is nothing quite like trying to put 270 DIN horses through tires 12 inches wide on 13 inch rims with a surface best described as slippery. Fifth was not a gear I needed much of.

Weather and pure cowardice aside, driving the "out-dated" 907 showed clearly how keen a factory must be which can retire such steeds. After all, it puts out more horsepower per liter than the 3-liter cars with a 123 figure.

The 907 is tractible as well, starting easily from a pit road so icy you could hardly stand up and thread a path through massed minions of the press. And it will pick up from 3500 revs if you really chicken out of a ball-bearing bend, yet howl up to 7500 before you can say "crossed arms on ice do not a happy pretzel make," one time fast. Yet in some ways it was easier to handle than their latest 911 R, largely because a GT shell gives the illusion that you know how.

This "R" designation is a catch-all meaning "hottest mill we can shoe-horn into more or less stock bodywork." While a works-tuned 911 S GT produces 180 injected hp and weighs nearly 2100 lbs, the R of today is almost 350 lbs slimmer, 10 mph quicker at 153 and blessed with 230-hp worth of 4-cam six. It too requires a light foot among the snow flakes. Study that spec! It's not the usual six with one cam over each bank but a new four-cammer pointing the path to 1975.

This Porsche outfit goes to the races, Dad. 1969 will be their 18th straight year on the big swing, never minding speed records of which Steinemann holds five. And they still keep up the gents-driver image. Would you believe every single driver got out of fire-proof overalls (arctic warm-up suits would have gone on better) and into not only three-button suits but matching vests before the speeches began? ●

This is the business end of the 908 spyder, Porsche's most potent though not fastest weapon for the 1969 season.

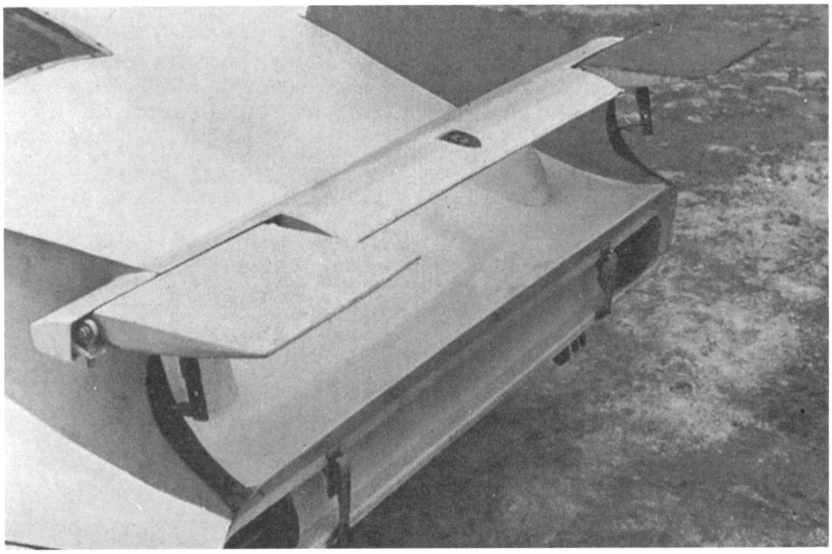

Long-tail version of the 908 has suspension-worked flippers with a fixed airfoil and fins, current epitome of Porsche aerodynamic thinking.

The 911 R with four cams and six injected cylinders is Porsche's hottest road car.

PORSCHE 910

ENGINE rear, 4 stroke; cylinders: 6, horizontally opposed; bore and stroke: 3.15 × 2.60 in, 80 × 66 mm; engine capacity: 121.15 cu in, 1,991 cu cm; compression ratio: 10.3; max power: 220 hp at 8,100 rpm; max torque: 152 lb ft, 21 kg m at 6,400 rpm; specific power: 110.5 hp/l; cylinder block: light alloy, chrome liners; cylinder head: light alloy, hemispherical combustion chambers; crankshaft bearings: 4; valves: 2 per cylinder, overhead, Vee-slanted, dual ignition; camshafts: 1, overhead per cylinder block; lubrication: gear pump, filter on by-pass, dry sump, oil cooler; lubr cating system capacity: 24.64 imp pt, 29.60 US pt, 14 l; carburation: Bosch indirect injection; fuel feed: electric pump; cooling system: air-cooled.

TRANSMISSION driving wheels: rear; clutch: single dry plate; gearbox: mechanical; gears: 5 + reverse; gear lever: to driver's right; final drive: spiral bevel; axle ratio: 4.420.

CHASSIS tubular trellis frame; front suspension: independent, wishbones, coil springs, anti-roll bar, telescopic dampers; rear suspension: independent, wishbones, torque arms, coil springs, anti-roll bar, telescopic dampers.

STEERING rack-and pinion ZF.

BRAKES disc, dual circuit.

ELECTRICAL EQUIPMENT voltage: 12 V; generator type: alternator, 490 W; ignition distributor: Bosch; headlamps: 4.

DIMENSIONS AND WEIGHT wheel base: 90.55 in, 2,300 mm; front track: 52.36 in, 1,330 mm; rear track: 55.20 in, 1,402 mm; overall length: 161.81 in, 4,110 mm; overall width: 66.14 in, 1,680 mm; overall height: 38.58 in, 980 mm; dry weight: 1,268 lb, 575 kg; tyres: front 5.25 × 13, rear 7.00 × 13; fuel tank capacity: 22 imp gal, 24.4 US gal, 100 l.

PERFORMANCE max speed: 174 mph, 280 km/h.

The Porsche is undoubtedly the most formidable car with a 2 litre engine capacity. Thanks, in fact, to its perfect preparation and to scrupulous, Teutonic thoroughness, the German car has managed to obtain a plentiful series of outstanding results. With the exception of a few occasions on which a variation with an 8 cylinder engine (engine capacity from 122.04 cu in, 2,000 cu cm, to 134.24 cu in, 2,200 cu cm) was raced, all efforts were concentrated on a single model of racing car, the 2000 prototype, and this policy gave excellent results. The engine, the usual air-cooled "boxer", is the direct descendant of the 91.53 cu in, 1,500 cu cm type built for the F 1 in 1962.
It is interesting to note that today Porsche is the only factory building air-cooled engines. The makers have unrivalled experience in this field and this allows them to obtain the same kind of power as with water-cooled engines but with the advantage of doing away with the surplus weight of the cooling system.
The coachwork which, like many of the car's accessories, is in fibreglass is a model of lightness and perfection.
The Porsche is the only make to adopt the Bosch system of indirect injection with separate plungers for each cylinder, very similar to the direct system.

PORSCHE 908 SPYDER

The 3-liter flat-8 that won the Manufacturers Championship for Stuttgart-Zuffenhausen

SPECIFICATIONS—PORSCHE 908 SPYDER

Engine . . . . flat-8, dohc
Bore x stroke, mm . . . . 85 x 66
Displacement, cc 2997
Compression ratio 10.4:1
Bhp @ rpm . . . . 350 DIN @ 8400
Torque @ rpm, lb-ft. . 235 DIN @ 6600
Carburetion . . . . . . Bosch fuel injection
Ignition . . . . double transistor
Lubrication . . . . dry sump
Transmission . . . . 5-sp all-synchro
Final drive ratio . . . . wide choice to suit different circuits
Frame . . . . multi-tube
Brakes . . . . disc, dual circuit
Wheels . . . . 15-in 5-spoke alloy
Tires . . . . Dunlop, 4.50/11.60-15 front, 5.50/13.60-15 rear
Steering . . . . rack and pinion
Front suspension: independent by upper and lower A-arms, coil springs, telescopic shock absorbers, anti-roll bar
Rear suspension: independent by single upper and lower arms, radius arms, coil springs, telescopic shock absorbers, anti-roll bar
Dry weight, lb 1385
Weight/power ratio, lb/bhp 3.9
Wheelbase, in 90.4
Track, front/rear 58.5/57.2
Length 157.4
Width 72.0
Height, top of bodywork 28.7
Engine oil capacity, qt 18.0
Fuel tank capacity, gal 31.7
Maximum speed, mph 174

Differing specifications for 908 Lang (Coupe):

Transmission . . . . 6-sp all-synchro
Tires . . . . 4.75/11.30-15 front, 6.00/13.50-15 rear
Dry weight, lb 1496
Weight/power ratio, lb/bhp 4.3
Length 190.5
Height 37.0
Maximum speed, mph 199

GEOFFREY GODDARD PHOTO

Long-tailed, better streamlined 908 Spyder ran at Le Mans, driven by Siffert/Redman, but retired with gearbox failure.

CUTAWAY DRAWING BY VITTORIO DEL BASSO

COURTESY POWERSLIDE

PORSCHE FACTORY GROUP 6 EFFORT
1969 INTERNATIONAL CHAMPIONSHIP FOR MAKES

te	Event	Entries	Finishers	Placings	Points
b. 1–2	Daytona 24-Hr	5 908 Coupes	0	None (All 5 cars retired with camshaft drive gear failure)	4 (Group 3)
ar. 22	Sebring 12-Hr	5 908 Spyders	3	3rd (Buzzetta/Ahrens), 5th (Mitter/Schutz), 7th (Elford/Attwood)	4
r. 13	Brands Hatch 6-Hr	4 908 Spyders	4	1st (Siffert/Redman), 2nd (Elford/Attwood), 3rd (Mitter/Schütz), 6th (Stommelen/Herrmann)	9
r. 25	Monza 1000-Km	4 908 Coupes	2	1st (Siffert/Redman), 2nd (Herrmann/Ahrens)	9
ıy 4	Targa Florio	6 908 Spyders	5	1st (Mitter/Schütz), 2nd (Elford/Maglioli), 3rd (Herrmann/Stommelen), 4th (Von Wendt/Kauhsen), 22nd (Larrousse/Lins)	9
ıy 11	Spa 1000-Km	3 908 Coupes	3	1st (Siffert/Redman), 3rd (Elford/Ahrens), 4th (Stommelen/Herrmann)	9
ıe 1	Nürburgring 1000-Km	6 908 Spyders	5	1st (Siffert/Redman), 2nd (Stommelen/Herrmann), 3rd (Elford/Ahrens), 4th (Lins/Attwood), 5th (Von Wendt/Kauhsen)	9
ıe 14–15	Le Mans 24-Hr	3 908 Coupes, 1 908 Spyder	1	2nd (Herrmann/Larrousse)	6
ıy 13	Watkins Glen 6-Hr	3 908 Spyders	3	1st (Siffert/Redman), 2nd (Elford/Attwood), 3rd (Lins/Buzzetta)	9
g. 10	Zeltweg 1000-Km	2 908 Spyders	2	5th (Lins/Larrousse), 6th (Von Wendt/Kauhsen) (Race won by Siffert/Ahrens 917 with Redman/Attwood 917 3rd)	9 Group 4)
	Totals:	42 Entries	28 Finishers	= 66 percent reliability	77 (Best 5=45)

PERFECTIONISTS IN TH

There was a time at Porsche when one man used to assemble a complete engine. It took him 25 hours. When he was done, he was allowed to hammer his initials into the crankcase.

Regrettably, a tradition that has passed. But it is one true to the legend created by the remarkable Doctors Ferdinand and Ferry Porsche . . . a father and son so alike in their technical views that today's Porsche is just the kind of car the old man might have produced.

Illustrated here is a draughtsman's cutaway of the impressive Porsche 907, a right-hand-drive version of the Porsche 910 equipped with a 2.2 litre 8 cylinder injection engine. Predictably, it is rear-engined, air-cooled, a faultless performer . . . all hallmarks of earlier Porsche designs such as the rampaging Auto-Unions of the 30's, and the Volkswagen, first produced in 1937 and modified only slightly since then.

Weighing only 1,325 lb ready for racing, the 907 develops over 270 bhp at 8,600 rpm to complete a power-to-weight equation that was responsible for a notable series of successes in 1968. 1st, 2nd and 3rd in the Daytona 24 hours and Targa Florio, 1st and 2nd at Sebring, Nurburgring, Paris and the Austrian G.P. Understandably, Porsche took off the Challenge Mondial de Vitesse et d'Endurance for 1968.

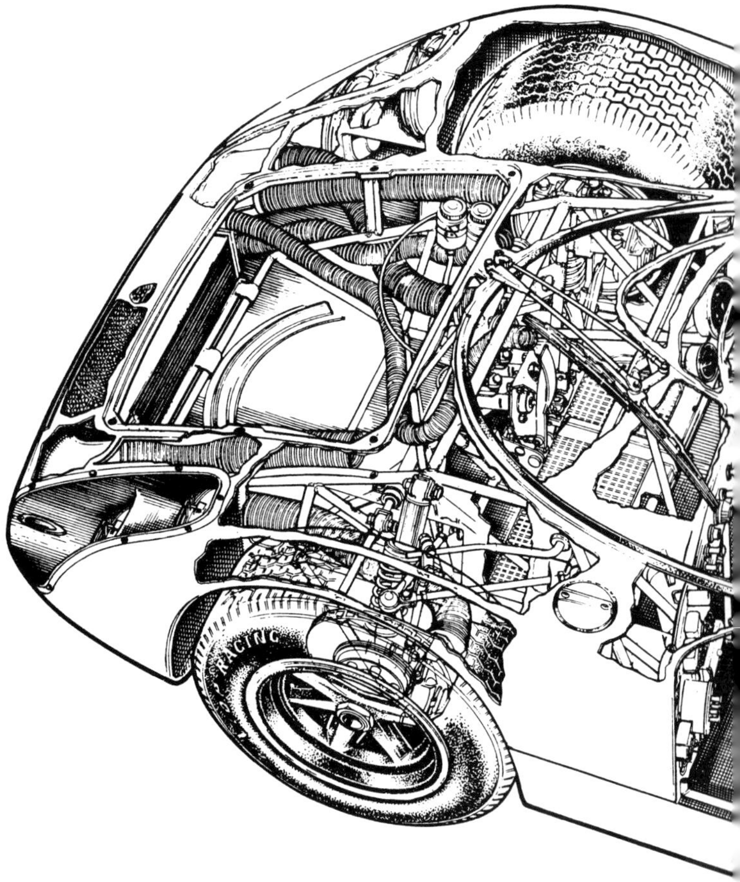

Facts about the Porsche 907

ENGINE: 2.2 litre Flat-8. COMPRESSION RATIO: 10.2-1. GEARBOX: 5-speed. Bosch Fuel Injection System. Bosch Transistorised Ignition and Alternator System. SUSPENSION: Koni dampers. Rack and Pinion Steering. BRAKING: Discs on all four wheels.

WHEELS: Light alloy. FUEL TANK: 22 gallons. BODY: Fibreglass. TYRES: 525 x 13 front; 700 x 13 rear. LENGTH OVERALL: 158.8″. WHEELBASE: 90.6″. TRACK: 47.6″ front; 45.2″ rear. GROUND CLEARANCE: 3.9″. WEIGHT: 1,325 lb ready for racing, without fuel or driver.

Shell an

In January, 1967
decision to do its
exclusively on Su
Benzine — and S
Thus, in the She
names as Ferrari,
and other marqu

IR ART: THE PORSCHES

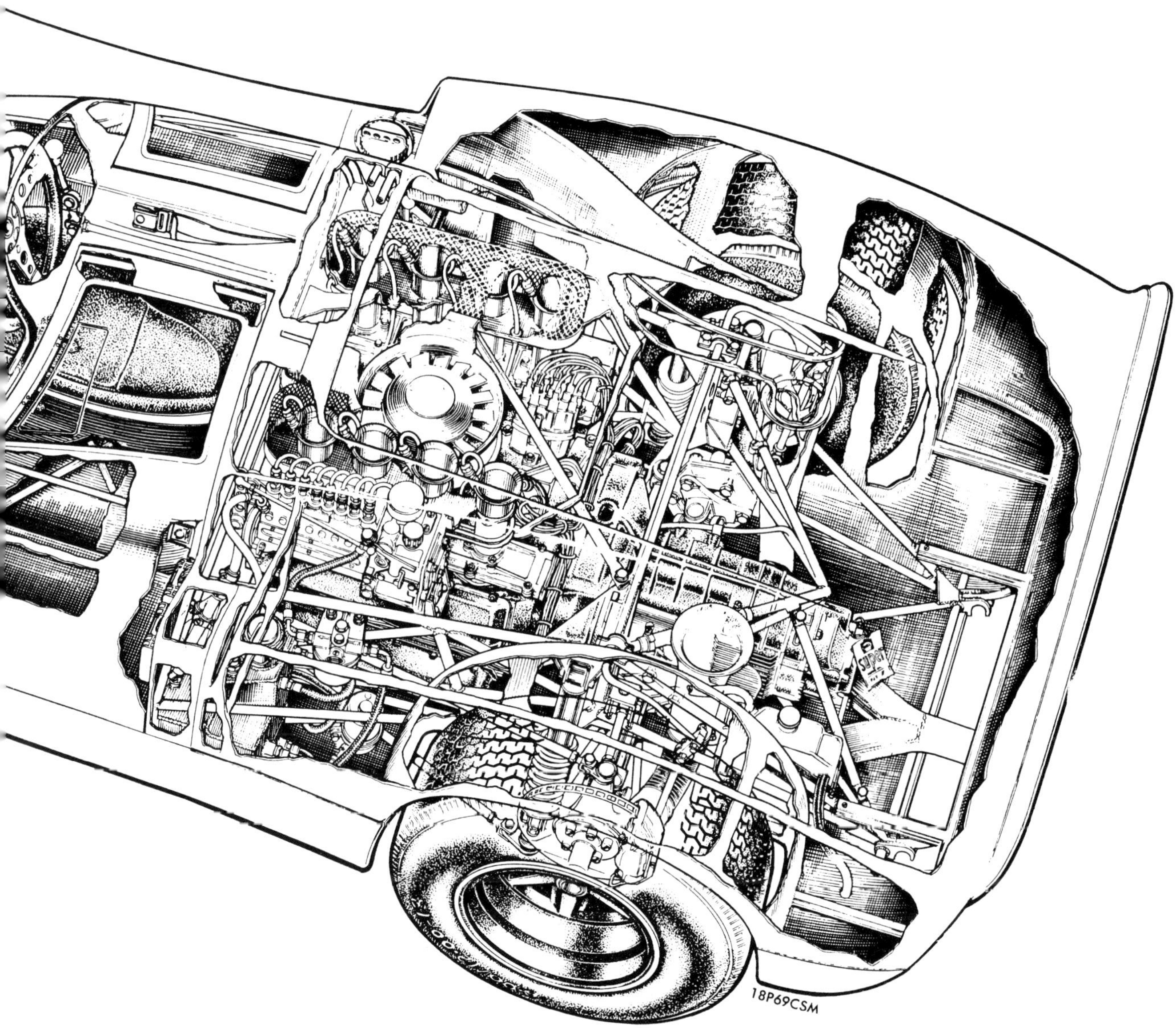

sche as partners

ıade the
ıg
ith Methyl
Motor Oil.
ɔined such
Chaparral

in long distance motor racing. Shell automotive experts have worked with Porsche engineers to achieve the outstandingly successful 'Go well' feeling of the last two years, not only for the 907 but for the 908 and 910 as well.

Go Well: Go → Shell

SHELL

Porsche Carrera 6

Michael Bowler tries one from the Stuttgart stable.

ANYONE around the International race-tracks of 1966 and for some time afterwards, will remember the 906 Carrera as one of the nicest looking of Porsche's generally dumpy-shaped cars. It had an urgency to its slender nose between high wheel arches that its more rounded predecessors lacked. Admittedly its back was familiarly round-shouldered but somehow the use of a full Plexiglass canopy, louvred down each side, masked the effect and the use of a spoiler across the chopped back provided further straightening. It might be thought that the 906 was born directly from the 904, the first of Porsche's mid-engined roadworthy cars – remember its dramatic 2nd overall performance in the hands of Bohringer in the 1965 Monte Carlo Rally? The 906 certainly used the fruits of experience with the 904 in its various forms, with four, six and eight-cylinder engines and as a spyder, but it was really a new concept for Porsche, sparked by the arrival of Ferdinand Piech. Where the late 'fifties cars had used space frames, the 904 had returned to a boxed punt to which the glassfibre body added stiffness. In spyder form with the flat-eight this had been inadequate to beat the Dino Ferrari on the European hill-climb scene, so the engineers had returned to the space frame in the quest for lightness; they also used 13-inch wheels and Lotus uprights for that car.

From the 904 came a wheelbase dimension plus suspension and brake components, but the track was wider and spherical joints were used for all the pick-up points where before wishbones had pivoted on rubber bushes. The configuration was similar therefore, with twin wishbones and coil springs fore and aft.

Porsche were to homologate the 906 as a group 4 car of the time, so at least 50 space frames had to be built and these were subcontracted out. With two of the tubes used to carry oil, inspection was important; perhaps Porsche learnt their theory of pressurised frame tubes on the 917 from discovering oil leaks in 906 frames!

The 904 was originally planned with a flat-six but this wasn't ready in time so they made do with the old Carrera 4 unit, but the 'six' was ready for the 906 and was being asked to turn out some 220bhp from 2-litres against the 130bhp that was the ration for the 911 of the time. The search for lower weight with the same durability produced the main barrel of the engine in magnesium alloy but the 911's Silumin heads, crankshaft and eight main bearings were retained. Titanium was used for connecting rods and various bolts. Extra power came from reworked heads with bigger ports and valves and a pair of triple-choke Weber 461 carburettors. The outcome was a claimed 210 at 8000rpm, as usual, a conservative claim. This was mated to a transmission similar to that of the 911 with magnesium castings again used.

All the development took place over the winter of 1965/6 and the cars were promised for delivery to the various racing-minded buyers for the start of the season. For Porsche that was Daytona in February and the first 906 came home sixth behind the Ford steamroller and a single Ferrari, the little two-litre on the tails of 7-litre V-8s and the long-proven strength of Maranello V-12s. By mid-April the required 50 had been built and the success of the 906 was on its way.

Outright victory came in the Targa Florio where Porsche had long held sway on the famous circuit where agility is at a premium. But to some extent they were hollow victories in the classes as there was no opposition – no one else was dedicated enough to lay down 50 racing cars at that stage. But they were always close to the unlimited capacity leaders finishing seventh at Le Mans in 1966 and 1967.

In England Bill Bradley won the RAC sports car championship with a 906 for 1968. By this time fuel injection had been homologated and 906 long-tails were 4th, 5th and 6th at Le Mans; in fact the injection only produced another 10bhp with a little more torque at even higher revs, now 152lb ft at 6400rpm, but the response was better than on carburettors. By now we were into the era of 908, 910 nearly 917, but the 906 was to be seen on the circuits for a few more years in the hands of privateers in long-distance races – it was a strong, fast and effective racer. Apart from the original 50 cars built for homologation in 1966, there were another two carburetted cars, nine fuel injected ones and four prototypes with flat-8s, so the numbers are limited, but there should be still many around the world.

The one shown here and in the centre spread was chassis 109, which has recently been restored for the American head of Anglo-American Automobile Investments, Sid Colberg, by the English partners. It came from Sweden and has had a stem-to-stern rebuild with effectively a whole new outer skin, screens and all. Body moulds came from Bill Bradley who has retained his Porsche interests since his 906 days. Bill Bradley Racing operates from Catesby Lane, Lapworth, Solihull, West Midlands and can supply parts for the production racers 904, 906 and 910 particularly as well as for the prototypes 907, 908 and the "production" 917 – mostly body parts for these but some mechanical parts as well. The Stuttgart end of his operation deals with Lolas but is manned by two 906 mechanics who retain the 906 jig over there.

The overall result of the restoration is to produce an as-new car; I'd prefer a different colour combination than black and aluminium but it's not my car! I had the chance to have a brief run at Silverstone where the car was having its post-rebuild shake-down; it wasn't quite right at that stage, with a fault in the oil circuit which put oil over the cockpit floor, but that has since been rectified. However it provided a sensible limiting effect at the time.

The car still seemed to have long circuit ratios, as it was very difficult to get away from rest on a high bottom gear and very little low speed torque, but, once on the move, it felt superb and initial reservations about left-hand-drive soon disappeared. Once over 5000rpm that lovely baying howl comes in with the power and the car really steps off. It was a bit nervous on the straight with the tyres nibbling at raised surfaces, but the steering provides plenty of feel and it does seem to be a very pleasant long-distance car. The thing that stands out is the solid Porsche-type integrity which seems to exist despite its very light overall weight; it feels like a racer, but it still feels as if it would keep going for a very long time. I didn't do enough laps to comment on the handling but one would expect it to be good and safe, a classic case of what looks right is right. ●

(Acknowledgement is made to *Porsche: Excellence was expected by* Karl Ludvigsen, published by Automobile Quarterly).

Above left, this interior shows a classic case of fitness for the purpose.

Above right, with the rear end removed the engine and transmission and suspension is easily worked on and the space frame chassis can be clearly seen.

Above, with the rear body panel back on, showing large rear window with engine vent slots cut in it. Far left, better looking, perhaps from this angle than from the back. Left, the Porsche six cylinder unit in all its glory.

PORSCHE 908

ENGINE rear, 4 stroke; cylinders: 8, horizontally opposed; bore and stroke: 3.35 × 2.60 in, 85 × 66 mm; engine capacity: 182.82 cu in, 2,996 cu cm; compression ratio: 10; max power: 310 hp at 8,400 rpm; specific power: 105 hp/l; cylinder block: light alloy, chrome rods; cylinder head: light alloy, hemispherical combustion chambers; crankshaft bearings: 10; valves: 2 per cylinder, overhead, Vee-slanted; camshafts: 2 per cylinder block, overhead; ignition: dual; lubrication: gear pump, dry sump, oil cooler; lubricating system capacity: 35.52 imp pt, 41.60 US pt, 18 l; carburation: Bosch indirect injection; fuel feed: electric pumps; cooling system: air-cooled, fan on crankshaft.

TRANSMISSION driving wheels: rear; clutch: dry multi-plates; gearbox: mechanical; gears: 6 + reverse; gear lever: side; final drive: bevel, limited slip.

CHASSIS tubular trellis with fibreglass body; front suspension: independent, wishbones, coil springs, telescopic dampers; rear suspension: independent, wishbones, coil springs, anti-roll bar, hydraulic telescopic dampers.

STEERING rack-and-pinion.

BRAKES disc, dual circuit.

ELECTRICAL EQUIPMENT voltage: 12 V; alternators: 2, one mounted direct on crankshaft, the other driven by left-hand semi-axle; distributors: Bosch; headlamps: 4.

DIMENSIONS AND WEIGHT wheel base: 90.55 in, 2,300 mm; front track: 58.27 in, 1,480 mm; rear track: 57.87 in, 1,470 mm; overall length: 183.07 in, 4,650 mm; overall width: 67.72 in, 1,720 mm; dry weight: 1,532 lb, 695 kg; tyres: 4.75/11.30 × 15 front, 6.00/13.50 × 15 rear; fuel tank capacity: 26.40 imp gal, 31.7 US gal, 120 l.

PERFORMANCE max speed: 198.8 mph, 320 km/h.

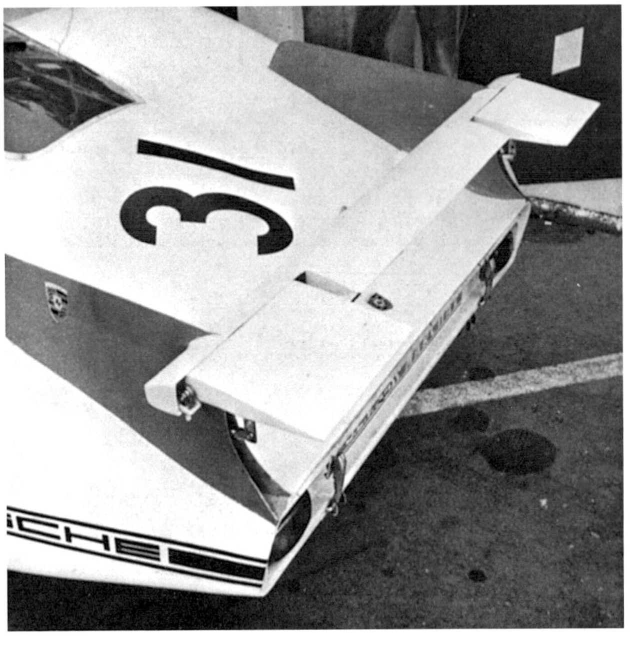

The 3-litre version of the Stuttgart works' racing car takes the place of the type 907 which has an engine capacity of 2.2 litres. The engine, however, has been completely redesigned, to take care of the increased distance between the cylinders and to allow a proper dimensioning of the parts.
The cooling system, too, has been modified and the solution of a fan mounted on the very end of the crankshaft beside the alternator has been used, with the aim of eliminating any possibility of trouble arising from the presence of a belt. However, on its first public appearance, the engine showed it was not yet quite ready for long-distance races. For the Le Mans 24 Heures, the car still had a long-tailed streamlined body, with two mobile fins connected to the suspension and moving so as to increase the aerodynamic pressure when the tail of the car tended to lift for any reason. The Porsche, which from many points of view is remarkable, is still the only air-cooled racing car.